CRAYFISHES, LOBSTERS AND CRABS OF EUROPE

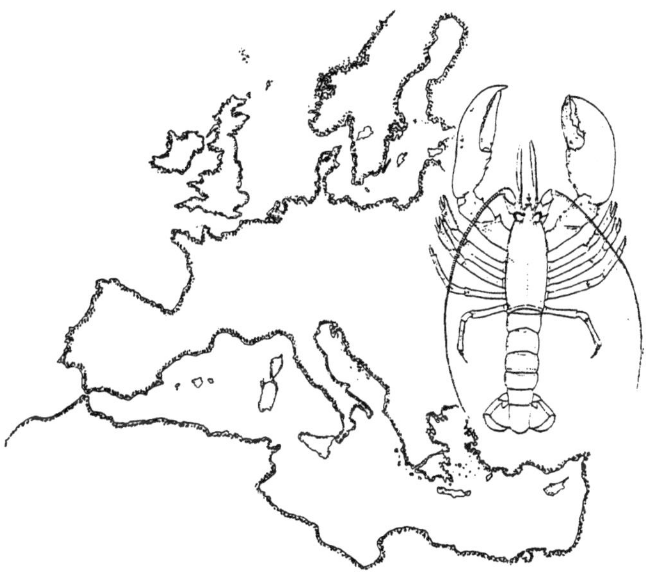

Fishing for the European Clawed Lobster *circa* 1800. From Bell, T. (1845) *A History of the British Crustacea*, p.140.

CRAYFISHES, LOBSTERS AND CRABS OF EUROPE

AN ILLUSTRATED GUIDE

to common and traded species

RAY INGLE

Former Senior Curator of Crustacea
The Natural History Museum, London

SPRINGER-SCIENCE+BUSINESS MEDIA, B.V.

First edition 1997

Originally published by Chapman & Hall in 1997
Softcover reprint of the hardcover 1st edition 1997

ISBN 978-94-010-6482-8 ISBN 978-94-011-5872-5 (eBook)
DOI 10.1007/978-94-011-5872-5

A catalogue record for this book is available from the British Library

Library of Congress Catalog Card Number: 97-68951

∞ Printed on acid-free text paper, manufactured in accordance with ANSI/NISO Z39.48-1992 (Permanence of Paper).

CONTENTS

Preface ix

1 Structure and function 1
- 1.1 External features 1
- 1.2 Exoskeleton 4
- 1.3 Internal features 4
- 1.4 Moulting 5
- 1.5 Sexing decapods 5

2 Classification and nomenclature 11
- 2.1 Taxonomy 11
- 2.2 Nomenclature 11
- 2.3 Vernacular names 12
- 2.4 Identifications 13
- 2.5 Changes in classifications 13
- 2.6 Systematic classification of decapods 13

3 Overview of groups 21
- 3.1 Freshwater crayfishes and clawed lobsters 21
- 3.2 Crawfishes and shovel-nosed lobsters 22
- 3.3 Mud lobsters 22
- 3.4 False lobsters and crabs 22
- 3.5 True crabs 23

4 Freshwater Crayfishes (Astacoidea) 27
- 4.1 Native species 27
- 4.2 Imported species 28
- 4.3 Life history 28
- 4.4 Crayfish farming 28
- 4.5 Identification 33

5	**Marine Clawed Lobsters (Nephropoidea)**	**41**
	5.1 Native species	41
	5.2 Commercially imported species	41
	5.3 Life history	42
	5.4 Fisheries for homarid lobsters	44
	5.5 Fisheries for nephropid lobsters	50
	5.6 Marketing clawed lobsters	51
	5.7 Aquaculture of homarid lobsters	53
	5.8 Identification	54
6	**Marine Crawfishes and Shovel-nosed Lobsters (Palinuroidea)**	**61**
	6.1 Native species	62
	6.2 Commercially imported crawfishes	62
	6.3 Fishing methods for crawfishes	62
	6.4 Life history	63
	6.5 Aquaculture of crawfishes	64
	6.6 Identification	64
7	**Mud Lobsters (Thalassinoidea)**	**75**
	7.1 Native species	75
	7.2 Life history	75
	7.3 Bionomics	76
	7.4 Identification	76
8	**Hermit and Stone Crabs (Paguroidea)**	**83**
	8.1 Native species of hermit crabs	83
	8.2 Life history of hermit crabs	83
	8.3 Bionomics of hermit crabs	84
	8.4 Identification of hermit crabs	85
	8.5 Native species of stone crabs	94
	8.6 Life history of stone crabs	94
	8.7 Commercial exploitation of stone crabs	95
	8.8 Identification of stone crabs	95
9	**Squat Lobsters and Porcelain Crabs (Galatheoidea)**	**99**
	9.1 Native species of squat lobsters	99
	9.2 Commercially imported squat lobsters	100

9.3 Life history of squat lobsters 100
9.4 Bionomics of squat lobsters 101
9.5 Identification of squat lobsters 101
9.6 Native species of porcelain crabs 108
9.7 Life history of porcelain crabs 108
9.8 Bionomics of porcelain crabs 109
9.9 Identifcation of porcelain crabs 109

10 True Crabs (Brachyura) 115
10.1 Native species 115
10.2 Introduced and immigrant species 115
10.3 Native species of economic importance 116
10.4 Life history 118
10.5 Life styles 119
10.6 Identification 120

11 Imported Crayfishes, Lobsters and Crabs 177
11.1 Freshwater crayfishes 178
11.2 Clawed lobsters 190
11.3 Crawfishes 198
11.4 Shovel-nosed lobsters 215
11.5 Squat lobsters 220
11.6 Stone crabs 222
11.7 True crabs 226

Colour plates 233

Index 267

Preface

The lobster served in a restaurant or displayed on a fishmonger's counter and the crabs disturbed while exploring intertidal rock pools are, for many, the first encounter with the large class of invertebrates-the Crustacea. However, professional fishermen, marine aquarists, biologists studying seashore and coastal waters, those involved in trading shellfish and even many restauranteurs are very aware that a great number of lobster and crab-like crustaceans inhabit the seas surrounding the British Isles, Northern Europe and the Mediterranean. Some of these, in addition to species imported from other countries, are marketed for human consumption.

This present book will enable the reader to identify the species of lobsters and crabs that are most frequently seen in the North Eastern Atlantic and Mediterranean regions, either during shore explorations, landed by fishing boats, displayed in public aquaria, in fish markets or available in restaurants. A few deep-water oceanic species are included as these are often of considerable local interest when occasionally brought ashore from vessels fishing deep waters. The species of freshwater crayfishes are included as these crustaceans are increasingly sought after delicacies and are frequently offered for sale in fish markets. A number of species of lobsters (or their 'tails') and a few crab species are imported into UK and European markets. These are included here for the benefit of shellfish importers, customs and excise officers regulating shellfish tariffs and students studying for fishery diplomas, all of whom often have to deal with the identifications of these exotic species.

There are in excess of two hundred and thirty species of lobster and crab-like crustaceans inhabiting north eastern Atlantic and Mediterranean waters. Forty two of these are described and illustrated in this book. For others and for more information about those included here, the reader will need to consult one or more of the technical monographs, guides or papers listed under 'Further reading'at the end of each chapter. These publications are available from specialist book sellers but many usually can be obtained on loan through the services provided by local libraries, or can be consulted, usually by prior arrangement, in libraries of some of the major museums.

Vernacular names are limited to those most commonly used and are listed for countries where the species described is likely to be most frequently seen or is traded. Many alternative vernacular names are cited in the the privately published book by G.N Cherry (see p. 20) and in FAO publications included in 'Further Reading' at the end of chapters. Abbreviations for terms used throughout this book are as follows.

Countries. An-Angola; Dn-Denmark; Fi-Finland; Fr-France; Ga-Gambia; Gh-Ghana; Gn-Germany; Gr-Greece; It-Italy; Mr-Morocco; Nd-Netherlands; Nw-Norway; Pt-Portugal; Sc-Scandinavia (Norway & Sweden); Sp-Spain; Sw-Sweden; Tn-Tunisia; Tr-Turkey; UK-United Kingdom.

Language. Gr-Greek, Lt-Latin, Ni-New Latin; Ti-Tupi Indian.

Morphology. **ab**-abdomen; **ae**-antennal spine; **al**-antero-lateral margin; **an1**-first antenna; **an2**-second antenna; **ap**-antennular plate; **ar**-carapace areole; **as**-antennal scale; **bas**-basis; **br**-branchial region; **ce**-chela [=propodus+dactylus of first pereiopod]; cheliped=**pr1**-first pereiopod; **cg**-cervical groove; **cl**-small claw; **cox**-coxa; **cp**-cephalothorax; **cr**-carapace; **crp**-carpus; **dcy**-dactylus; **dg**-dorsal groove; **dm**-dorsomedial ridge; **en**-endopod; **ex**-exopod; **ey**-eye; **fg**-female's genital opening; **fh**-frontal horns; **fl**-flagellum; **fr**-frontal region; **gr**-grooves (dorsomedial on abdomen); **im**-intersegmental membrane; **isc**-ischium; **it**-involuted part; **lb**-post-frontal lobe; **li**-lateral incision; **lm**-lateral margins; **lp**-longitudinal process; **ls**-longitudinal suture; **lt**-lateral teeth; **md**-mandible; **mg**-male's genital opening; **mr**-median ridge; **mrs**-merus; **mv**-medial groove; **mxpd1** to **3**-first to third maxillipeds; **mx1** and **2**-first and second maxillae; **ob**-orbit; **oc**-ocular acicle; **om**-outer margin; **pc**-post-frontal crest; **pd**-postrostral ridge; **pe**-peduncle; **pl**-pleopod; **pl1**-modified first pleopod [gonopod] of male; **pn**-protuberance; **pr1** to **pr5**-first to fifth pereiopods; **pro**-propodus; **ps**-postero-lateral margin; **pt**-posterior tooth; **pu**-pleuron; **rp**-female's sperm receptacle; **rs**-rostrum; **sc**-scaphognathite; **sh**-shield; **sm**-submedial spines; **sp**-spine; **spr**-supraorbital spines; **st**-sternum; **td**-transverse ridge; **tg**-transverse groove; **tl**-telson; **tr**-tergite ridge; **ur**-uropod.

Acknowledgements. The majority of lobsters and crabs illustrated are from specimens in the collections of the Natural History Museum, London and I thank Paul Clark of the Crustacea Section for providing laboratory bench space and for his assistance on various matters. For details of some species of Spiny Crawfishes, Rock Crawfishes, Shovel-nosed lobsters, in addition to Museum specimens, I have have been guided by Dr Lipke Holthuis's highly informative FAO publication on Marine Lobsters of the World (see p. 60 for full citation) and to whom I give full acknowledgement for help received.

I also wish to thank Dr Eric Edwards, OBE, The Shellfish Association of Great Britain, for providing current information on Edible Crab legislation and Dr David Bennett, MAFF Fisheries Laboratory, Lowestoft for European Council regulations on the fisheries for Norway Lobsters.

Fig. 7.2, showing generalised tunnels of mud lobsters is adapted from fig. 2 of the paper by Nash *et al* cited on p. 82.

I am grateful to colleagues in Customs and Excise for advice on Tariffs relating to shellfish.

Much of the information presented here on traded lobsters and crabs formed a part of seminars given to students from many countries attending the Overseas Development Administration *Diploma in Fisheries Management* course held in the School of Fisheries Studies, Humberside College of Higher Education. I am grateful for details about local fisheries that these students were sometimes able to provide and I hope this book will be of use to those involved in fisheries management who require a working knowledge about the taxonomy of lobsters and crabs .

1

Structure and function

The crustaceans described in this book are assigned to the Order Decapoda (Gr. *deca,* ten; *podotes*, footed). Decapods have armour-like body segments, some bearing appendages, held together by membranous and muscular tissue. The external features of a clawed lobster are described below and compared with those of a hermit and true crab.

1.1 EXTERNAL FEATURES

1.1.1 Clawed lobster (figs 1.1, 1.2)

Two major divisions of the body can be recognised: the head and thorax composing the *cephalothorax* (**cp**), and the segmented *abdomen* (**ab**). The lobster's body comprises nineteen segments each with a pair of limbs. The *eyes* (**ey**) and 'tail' or *telson* (**tl**) are not considered as true segments since they are formed during early development and before body segmentation begins. The segments of the cephalothorax are concealed by a shield-like outgrowth, the *carapace* (**cr**), but their positions can be established by noting from where the associated paired limbs arise. The carapace is extended downwards on both sides concealing and protecting the delicate gills; anteriorly it is extended into a *rostrum* (**rs**) that protects the eyes.

The first and second pairs of limbs are the *first* (**an1**) and *second antennae* (**an2**) bearing filamentous *flagella* (**fl**) used for "tasting" the water. The third to fifth pairs of limbs form the anterior appendages of the mouthparts and are best seen in 'head-on' view since they are usually only exposed when the lobster feeds. The third pair are the 'jaws' or *mandibles* (**md**) and along with the fourth pair, *first maxillae* (**mx1**) shred food. The fifth pair, *second maxillae* (**mx2**) also dismember food but their chief feature is the large paddle-shaped 'bailer' or *scaphognathite* (**sc**) that is periodically vibrated to drive water over the gill surfaces.

It is convenient to divide the following six to thirteen conspicuous pairs of thoracic limbs into two groups. Pairs six to eight are the *first* to *third maxillipeds* (**mxp 1-3**). They form the posterior appendages of the mouthparts and assist with holding and dismembering food. Limbs on segments nine to thirteen are the *first* to *fifth pereiopods* (**pr1-pr5**), appropriately named 'walking legs'. The first to third pair terminate as pincers (chelae). The first pair are often termed *chelipeds* and the large chelae are used for defence, mating and sometimes for holding food. The second and third pairs of pereiopods have smaller chelae that are used to pass food forward to the mouthparts and to preen the body and limbs. The fourth to fifth pairs of pereiopods terminate in sharp claws. When the lobster walks, using these pereiopods, the chelipeds are usually held out in front and the abdomen straightened.

All thoracic limbs are clearly segmented and these segments are joined to each other by thin intersegmental tissue. Some segments freely articulate with their neighbour although the plane of movement is usually limited. The limb segments are variously named as shown for the third maxilliped in fig. 1.2.

The *female's genital openings* (**fg**), through which eggs are extruded during spawning, open on the coxal segment of the fourth pair of pereiopods. The *male's genital openings* (**mg**), through which spermatophores issue during mating, are located on the coxae of the fifth pereiopods.

The *abdomen* (**ab**) is composed of six shelly 'rings' representing segments fourteen to nineteen. Each slightly overlaps the one following and they are joined by thin *intersegmental membranes* (**im**). Although each has a limited vertical articulation with its neighbour the abdomen has nevertheless, considerable flexibility particularly when it is 'flapped' when the lobster needs to perform a rapid backward escape movement. Ventrally paired limbs, *pleopods* (**pl**) arise from each segment. The pair on the first segment of the female's abdomen are small, but for the male these and the pair on the following segment (**pl1, pl2,**) are modified as *gonopods* for transferring spermatophores during mating, to the female's *sperm receptacle* (**rp**) where they are retained until the eggs are spawned. The second to fifth pleopods of the female and third to fifth of the male are similar in both sexes. They are two-branched paddle-shaped structures fringed with hairs [setae] and often termed "swimmerets". They have a limited respiratory function and in the female the developing eggs are attached to special *oosetae* (**os**) on these pleopods. Limbs of the sixth segment are stout, broad and flattened *uropods* (**ur**) and with the terminal *telson* (**tl**) form a broad 'tail fan'.

The sides of the carapace are the outer walls of a branchial chamber in which the gills are located, the lobster's chief organs of respiration. Gills arise from the side walls of the thorax above the limbs [pleurobranch gills], from the soft membranous tissues between limbs and body walls [arthrobranch gills], and from the coxal segments of limbs podobranch gills]. The gills are composed of a central rod encircled by attached rows of leaf-like processes, but filamentous forms of gills occur on or near blade-like structures attached to the proximal parts of coxal segments of the first to third maxillipeds and first to fourth pereiopods.

The external surfaces of the body, including front and hind parts of the gut are covered with a tough shell variously termed exoskeleton, integument or cuticle (p. 4). The exoskeleton is thickest in the carapace region and thin on the underside of the abdomen and the parts where limbs and abdominal segments articulate [the intersegmental membranes].

1.1.2 Hermit crabs (fig. 1.3, A & B)

The carapace of hermit crabs offers limited protection to the gills. Anteriorly it is hardened [calcified] as a *shield* (**sh**) but is membranous and soft posteriorly. The abdomen of hermits is twisted for fitting into the whorls of empty gastropod [marine snail] shells The *pleopods* (**pl**) are reduced in size and sometimes in number and the *uropods* (**ur**) are modified for bracing against the internal surface of the shell.

1.1.3 True crabs (fig. 1.3, C-F)

The carapace of true crabs is usually broader than long, a feature associated with their sideways mode of walking. The carapace is well calcified throughout, extends ventrally inwards to the limb bases, and completely covers the gills. The *sternum* (**st**) and sutures of the posterior sternites are prominent. The latter are partly covered by the flap-like *abdomen* (**ab**); the ventral surface of this bears the pleopods. The abdomen of the male is triangular (fig. D) whereas that of the female is broadly oval (fig. E). The female has two-branched setosed pleopods on the second to fifth abdominal segment but the male has pleopods [gonopods], modified for sperm transference (fig. F), developed only on the first and second segments.

1.2 EXOSKELETON

The exoskeleton of lobsters and crabs is composed of several layers. The outermost [epicuticle] is very thin except in regions subjected to abrasion, eg. tips of claws, legs and margins of articulating surfaces. The epicuticle is comparatively flexible because it is only lightly impregnated with calcium salts and does not contain chitin, a nitrogenous carbohydrate. The layer beneath is the endocuticle. It is composed of chitin and the upper two thirds is impregnated with calcium salts; combined, these give the exoskeleton its maximum hardness and strength. The uppermost part of the endocuticle contains granular deposits of pigment [pigmented layer] that gives lobsters and most adult crabs their permanent colour.

Some pores on the epicuticle surface are the openings of tegumental glands whose secretions produce the epicuticle. Other small pores mark the sites of possible chemoreception. Various microscopic hair-like structures [setae] project from the epicuticle. Some have known sensory functions detecting directions of water currents and sensing water disturbances signaling the approach of another animal.

1.3 INTERNAL FEATURES

The internal surface of the body shell of lobsters and crabs has numerous hard projections. These processes form a complex internal framework that provide additional strength to the exoskeleton, attachment for muscles, and support for some internal organs. This is termed the endophragmal skeleton and is particularly well developed near the basal parts of the cephalothoracic limbs where it forms a series of strong plates to accommodate end attachments of the powerful leg muscles. The sizes of muscles used to move body segments and limbs are related to function. For example, the powerful grip exerted by the larger of the paired chelipeds is achieved by the contraction of a powerful closer muscle that pulls the pivoted dactylus downwards to close the claw. Contraction of a smaller opener muscle as the closer relaxes causes the claw to open.

The best edible meat of the lobster is provided by the muscles in claws, legs and abdomen. These, and parts of the endophragmal skeleton, can be easily seen when removing meat from a cooked lobster.

1.4 MOULTING

Crustaceans grow by periodically moulting their hard exoskeleton that allows the expansion and subsequent hardening of a new one formed beneath the old. The exoskeleton of a recently moulted decapod is very soft, (crabs in this condition are termed 'peelers'). Moulting is a traumatic process and frequently occurs at night. For short periods before and after the event normal behaviour is temporarily suspended. Just prior to moulting water is absorbed causing the body to swell. The lobster rolls onto its side before moulting and a split occurs in the membrane between the dorsal part of the cephalothorax and abdomen. Through this tear the appendages and then the abdomen are carefully withdrawn from the old shell. Failure to successfully withdraw one or more limbs causes these to be shed and possible death due to excessive loss of blood. A lobster's new soft shell takes from twenty four to forty eight hours to harden, but for some crabs (North Atlantic Shore crab, for example) this process can take as long as sixteen days. After the shell has hardened normal life style is resumed.

Moulting is also an important part of the reproductive cycle. The lobster and most crabs are only receptive to mating when their shell is soft. Females exudes a chemical [pheromone] that attracts males and mate only during this soft-shelled stage. The males of some crabs, particularly swimming and European Edible Crabs, carry the female for a short time before she is due to moult and may even assist her with this process in order to mate immediately after her old shell has been discarded; the male will carry his partner until her shell has hardened.

1.5 SEXING DECAPODS

Sexing mature decapods is usually not too difficult. The position of the genital openings as shown for male (**mg,** fig. 1.1, C) and female (**fg,** fig. 1.1, D), are in the same positions for all groups and is the most reliable feature for determining the sexes of hermit crabs (fig. 1.3, B). Males of clawed and false lobsters have the first (less conspicuously in false lobsters, see p. 22) pair of pleopods highly modified as tubular *gonopods* (**pl1,** fig.

1.1, C). Males of crawfishes and mud lobsters have small rod-shaped appendages on the first pair of pleopods. Males of true crabs have the first and second pairs of pleopods styliform shaped (fig. 1.3, F). Males of true crabs are easy to recognise by the narrow triangular-shaped abdomen (fig. 1.3, D); the abdomen of the female is broad (fig. 1.3, E).

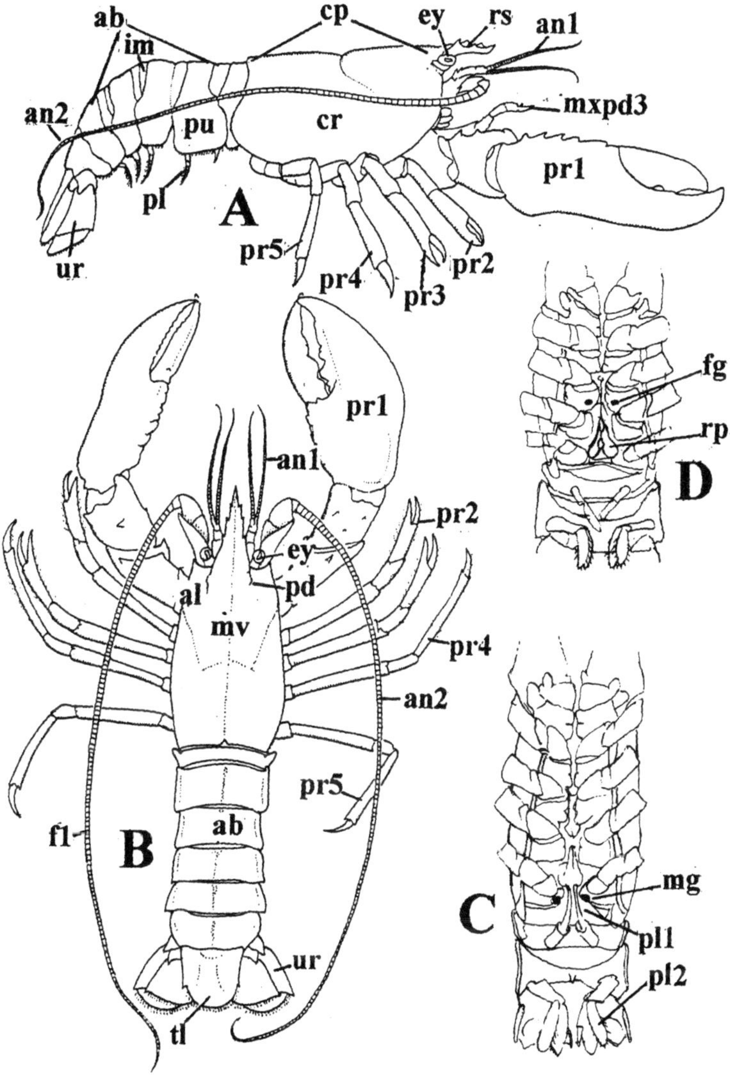

Fig. 1.1 External structure of a lobster-shaped decapod crustacean. A. Right lateral aspect. B. Dorsal aspect. Ventral aspects of cephalothorax of C-male and D-female: **ab** abdomen; **al** antero-lateral margin; **an1** first antenna; **an2** second antenna; **cp** cephalothorax; **cr** carapace; **ey** eye; **fg** female's left genital opening; **fl** flagellum of second antenna; **im** intersegmental membrane; **mg** male's left genital opening; **mv** medial groove; **mxp3** third maxilliped of right side; **pd** post-rostral ridge; **pl** pleopod; **pl1, pl2** male's modified first and second left pleopod [gonopods]; **pr1-pr5** first to fifth pereiopods; **pu** pleuron (side plate); **rp** female's sperm recaptacle; **rs** rostrum; **tl** telson; **ur** uropod.

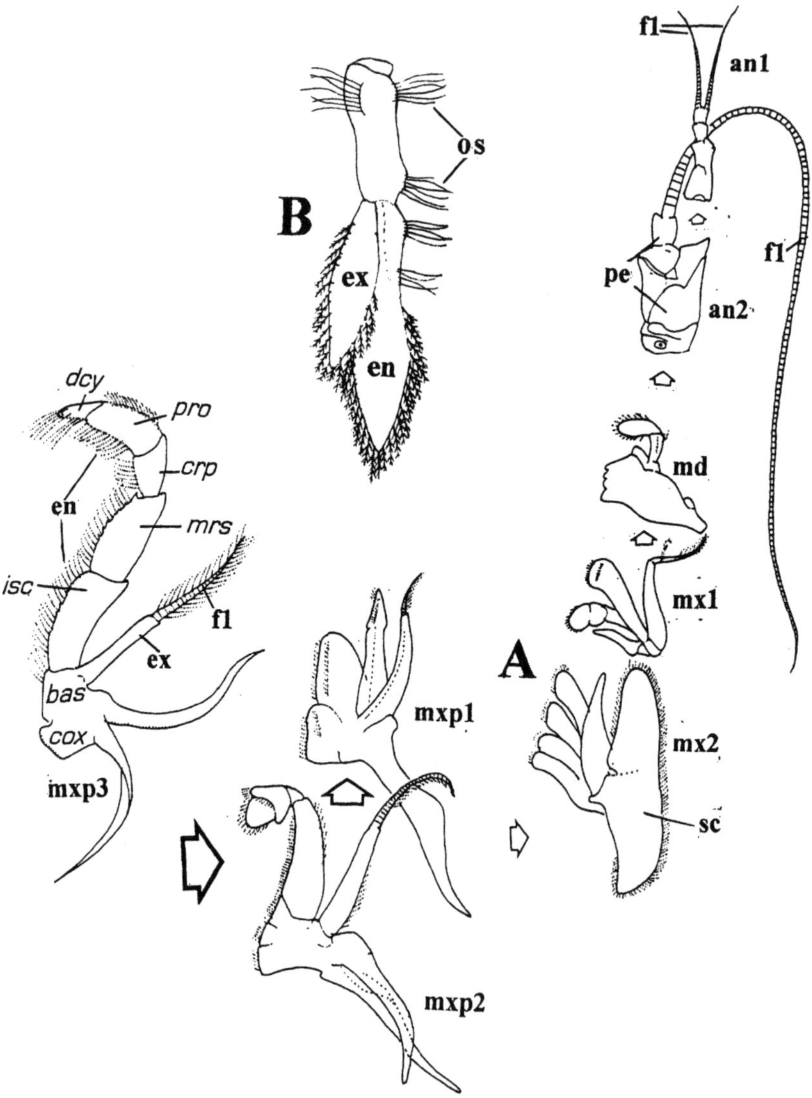

Fig. 1.2 A. Mouthparts and antennae of lobster removed from body from outer to inner order, as indicated by direction of arrows. B. Third right pleopod of female lobster: **an1** first antenna; **an2** second antenna; **en** endopod; **ex** exopod; **fl** flagella; **md** mandible; **mx1** & **2** first & second maxillipeds; **mxp1-3** first to third maxillipeds; **os** oosetae; **pe** peduncle; **sc** bailer or scaphognathite. Limb segments are named thus: ***bas*** basis; ***cox*** coxa; ***crp*** carpus; ***dcy*** dactylus; ***isc*** ischium; ***mrs*** merus; ***pro*** propodus.

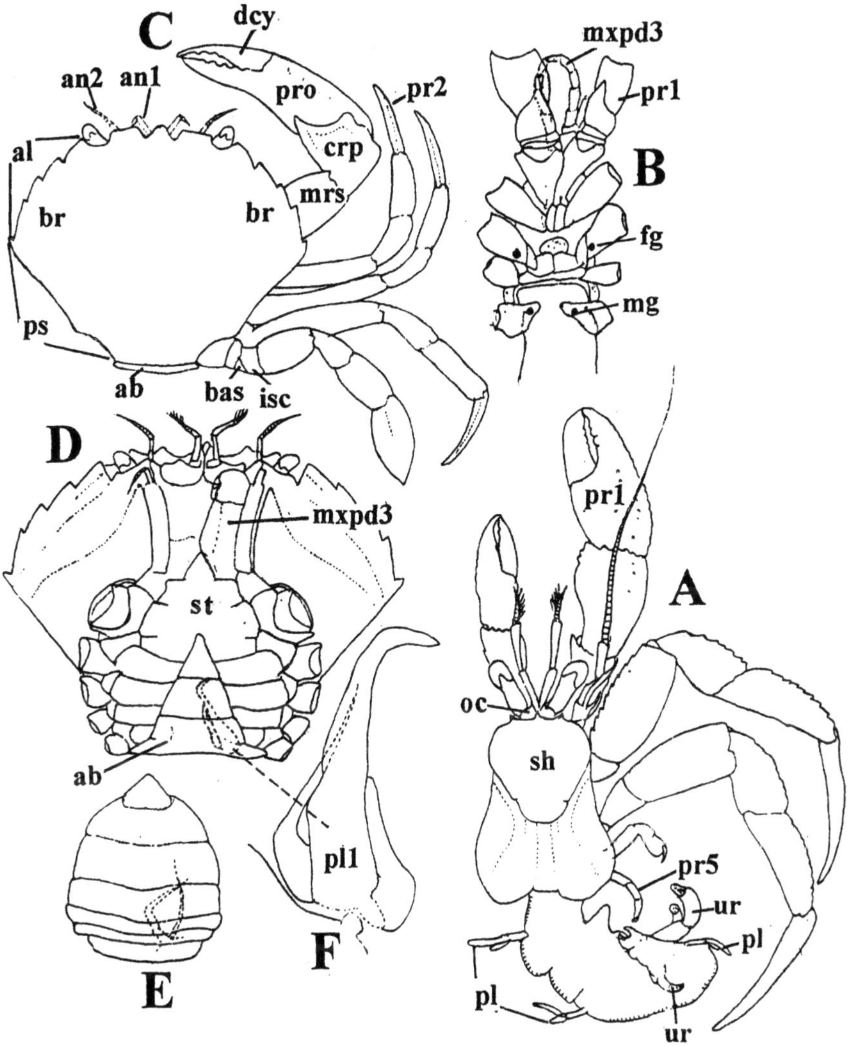

Fig. 1.3 External structure of a hermit crab.and true crab. A. Hermit crab, from dorsal aspect (pereiopods 2-5 from left side omitted). B. Hermit crab, ventral view of cephalothorax showing basal parts of pereiopods and positions of genital openings of female (on third) and of male (on fifth pereiopod). C. True crab, dorsal aspect of carapace and limbs on right side. D. True crab, ventral aspect to show sternum and folded abdomen. E. Abdomen of a female true crab from ventral aspect showing second left pleopod *in situ* (viewed as if the abdomen is transparent). F. First and second pleopods of male true crab from ventral view. Annotations as in figs 1.1 & 1.2 except for: **br** branchial regions; **oc** ocular acicles; **sh** shield; **st** sternum.

FURTHER READING

Lockwood, A.P.M., (1968) *Aspects of the Physiology of Crustacea.* Oliver & Boyde, London.

McLaughlin, P.A., (1980) *Comparative Morphology of Recent Crustacea.* Freeman & Co, San Francisco.

Warner, G.F., (1977) *The Biology of Crabs.* Elek (Scientific Books), London

Waterman, T.H. (Ed)., (1960) *The Physiology of Crustacea. Volume 1. Metabolism and Growth.* Academic Press, New York & London.

Waterman, T.H. (Ed)., (1961) *The Physiology of Crustacea. Volume II. Sense organs, Integration, and Behaviour.* Academic Press, New York & London.

2

Classification and nomenclature

2.1 TAXONOMY

The practice of classifying and naming organisms is termed taxonomy. Similar to other animals, crustaceans are classified on structural affinities. Some of these shared features are not always obvious but, in general terms, taxonomists look for structures that allow the construction of hierarchies in which shared characters become more restricted as organisms are arranged in a series of subordinate grades. These graded subdivisions are given various names, ie infraorder, section, superfamily, family, subfamily, genus, species and subspecies, reflecting the decreasing magnitude of shared features. The classification given on pp 13-19 for the crustaceans described in this book follows this arrangement.

Taxonomic characters are also used to construct keys highlighting the features that enable crustaceans to be assigned to the correct family, genus and species; keys of this form are usually included in many identification guides.

2.2 NOMENCLATURE

The study of scientific names and their application to organisms is termed nomenclature. The system used by taxonomists is based upon the one proposed by the Swedish naturalist Carl von Linnaeus and is applied to names published from 1758 onwards. The scientific name given to a species and subspecies are always latinized and usually printed in italic type. For example, the full citation for the European Clawed Lobster is: *Homarus gammarus* (Linnaeus, 1758). This comprises the generic name *Homarus*, the specific or trivial name *gammarus,* followed by the name of the author, in this instance Linnaeus who first described and named this lobster, then the earliest date (1758) when the name was published. Linnaeus used the name *Astacus gammarus,* but because the species was later reassigned to the genus *Homarus* Linnaeus's name is placed in brackets. Thus bracketed authors names always denote that the current scientific name cited is not that of the original combination.

2.3 VERNACULAR NAMES AND CLASSIFICATION

When available vernacular or 'common' names are often used for lobsters and crabs. This is quite acceptable for easily identified and common species for example, the North Atlantic Shore Crab *Carcinus maenas* (Linnaeus, 1758) and the European Edible Crab *Cancer pagurus* Linnaeus, 1758. However, confusion can arise when several alternative vernacular names, are frequently used, as is the case for some species.

In the USA attempts have been made to restrict vernacular names to particular decapod species and in Europe there are vernacular ' FAO recommended names' for fished species of lobsters and crabs. The restriction to only one vernacular name in all the major languages for (at least) each of the traded native and imported species of lobsters and crabs would considerably assist those involved in marketing shellfish and officials dealing with importation tariffs. In this respect the use of vernacular names at family level would also seem appropriate and helpful. Suggested names restricted to families and species are given below and further discussed in chapter 11.

- **Freshwater crayfishes** (all species of the superfamily Astacoidea).
- **Clawed lobsters** (all species of the superfamily Nephropoidea).
- **Crawfishes** (all species of the family Palinuridae).
- **Shovel-nosed Lobsters** (all species of the family Scyllaridae).
- **Mud lobsters** (all species of the infraorder Thalassinidea).
- **Hermit crabs** (all species of the families Diogenidae, Paguridae, Parapaguridae and Coenobitidae-species belonging to the last two mentioned families are not included in this book).
- **Stone crabs** (all species of the family Lithodidae).
- **Squat lobsters** (all species of the family Galatheidae).
- **Porcelain crabs** (all species of the family Porcellanidae).
- **True crabs** (all species of the Infraorder Brachyura).

2.4 IDENTIFICATIONS

Accurate identification and application of the correct scientific name is essential when the biology of a particular species is being studied or is subjected to legislation. For example, the UK Food Labelling Regulations of 1984 Sch I allows no other lobster than *Nephrops norvegicus* (Linnaeus, 1758) to be sold under the name of 'Scampi'. Also, it is now an offence to catch or sell specimens of our native White-foot Crayfish *Austropotamobius pallipes* (Lereboullet, 1858) that is protected by the Wildlife and Countryside Act of 1986.

2.5 CHANGES IN CLASSIFICATIONS

As our knowledge of taxonomy and nomenclature increases it is necessary to make changes in the classifications of organisms or to give priority to older names for species. Although these changes can sometimes confuse non-specialist they are usually indicated in most identification guides as synonyms at the beginning of a taxonomic account of a species; this practice is followed here.

2.6 SYSTEMATIC CLASSIFICATION OF DECAPODS

All species described or mentioned in this book arc listed below with a suggested English universal vernacular name. Bold-faced type indicate illustrated species. An* denotes that the species is an immigrant or has been introduced into the native fauna. A † indicates that the species is or may be traded in the UK or Europe.

INFRAORDER ASTACIDEA
Superfamily Astacoidea [Freshwater Crayfishes]
Family Astacidae
†***Astacus astacus astacus*** (Linnaeus, 1758) [Noble Crayfish]
Astacus astacus balcanicus (Karaman, 1929)
Astacus astacus colchicus Kessler, 1876
Astacus leptodactylus caspicus Eichwald, 1838
Astacus leptodactylus cubanicus Birstein & Vinogradov, 1934

Astacus leptodactylus eichwaldi Bott, 1950
†***Astacus leptodactylus leptodactylus*** (Eschscholtz, 1823) [Long-clawed Crayfish]
†***Astacus leptodactylus salinus*** Nordmann, 1842 [Long-clawed Crayfish]
Astacus pachypus (Rathke,1837) [Caspian Crayfish]
Austropotamobius fluviatilis lusitanicus Mateus, 1934
Austropotamobius italicus carsicus Karaman, 1962
Austropotamobius italicus italicus Karaman, 1962
†***Austropotamobius pallipes*** (Lereboullet, 1858) [White-clawed Crayfish]
Austropotamobius pallipes fulcisianus Nini, 1886
Austropotamobius pallipes italicus (Faxon, 1914)
Austropotamobius pallipes lusitanicus (Mateus, 1934)
Austropotamobius pallipes pallipes Bott, 1950
Austropotamobius torrentium (Schrank, 1803) [Stone Crayfish]
Austropotamobius (Atlantoastacus) bernhauseri Bott, 1972
*†***Pacifastacus (Pacifastacus) leniusculus leniusculus*** (Dana, 1852) [Signal Crayfish]
Family Cambaridae
*†***Orconectes (Faxonius) limosus*** (Rafinesque, 1817) [Striped Crayfish]
*†***Procambarus (Scapulicambarus) clarkii*** (Girard, 1852) [Red Swamp Crayfish]
Superfamily Parastacoidea
Family Parastacidae
Astacopsis gouldi Clark, 1936 [Tasmanian Crayfish]
Cherax albidus Clark, 1951 [Nurrabeil Yabby]
†***Cherax destructor*** Clark, 1936 [Melbourne Yabby]
†***Cherax tenuimanus*** Smith, 1912 [Marron]
Euastacus armatus (von Martens, 1866) [Murray Cray]
Superfamily Nephropoidea
Family Nephropidae [Clawed Lobsters]
Subfamily Nephropinae
†***Homarus americanus*** (H. Milne Edwards, 1837) [American Clawed Lobster]
†***Homarus gammarus*** (Linnaeus, 1758) [European Clawed Lobster]
†***Metanephrops andamenicus*** (Wood Mason, 1892) [Andaman Clawed Lobster]
Metanephrops armatus (Chan & Yu, 1991) [Armoured Clawed Lobster]
Metanephrops formosanus Chan & Yu, 1987) [Formosan Clawed Lobster]
Metanephrops japonicus (Tapparone-Canefri, 1873) [Japanese Clawed Lobster]

†*Metanephrops mozambicus* Macpherson, 1990 [African Clawed Lobster]
Metanephrops sagamiensis (Parisi, 1917) [Sculptured Clawed Lobster]
Metanephrops thomsoni (Bate, 1888) [Red-banded Clawed Lobster]
†*Metanephrops velutinus* Chan & Yu, 1991 [Velvet Clawed Lobster]
†***Nephrops norvegicus*** (Linnaeus, 1758) [Norway Clawed Lobster]
Subfamily Thymopinae
Nephropsis atlantica Norman, 1882 [Scarlet Clawed Lobster]
INFRAORDER PALINURA
Superfamily Palinuroidea
Family Palinuridae [Crawfishes]
Jasus (Sagmariasus) edwardsii (Hutton, 1875) [Red Rock Crawfish]
†***Jasus (Jasus) lalandii*** (H. Milne Edwards, 1837) [Cape Rock Crawfish]
Jasus (Jasus) paulensis (Heller, 1862) [St Paul Rock Crawfish]
Jasus (Jasus) tristani Holthuis, 1963 [Tristan Rock Crawfish]
Jasus (Sagmariasus) verreauxi (H. Milne Edwards, 1851) [Green Rock Crawfish]
†***Palinurus elephas*** (Fabricius, 1787) [European Spiny Crawfish]
†***Palinurus mauritanicus*** (Gruvel, 1911) [Atlantic Pink Spiny Crawfish]
†***Panulirus argus*** (Latreille, 1894) [Caribbean Spiny Crawfish]
†***Panulirus cygnus*** George, 1962 [Australian Spiny Crawfish]
†***Panulirus homarus*** (Linnaeus, 1758) [Scalloped Spiny Crawfish]
†***Panulirus ornatus*** (Fabricius, 1798) [Ornate Spiny Crawfish]
†***Panulirus polyphagus*** (Herbst, 1793) [Mud Spiny Crawfish]
†***Panulirus regius*** (de Brito Capello, 1864) [Royal Spiny Crawfish]
†***Panulirus versicolor*** (Latreille, 1804) [Painted Spiny Crawfish]
†***Puerulus sewelli*** Ramadan, 1938 [Arabian Whip Crawfish]
Family Scyllaridae [Shovel-nosed Lobsters]
Subfamily Arctinae [Spanish and Slipper Lobsters]
Arctides regalis Holthuis, 1963 [Royal Spanish Lobster]
Scyllarides herklotsii (Herklots, 1851) [Red Slipper Lobster]
†***Scyllarides latus*** (Latreille, 1803) [Mediterranean Slipper Lobster]
Subfamily Ibacinae [Fan and Mitten Lobsters]
Evibacus princeps S.I. Smith,1869 [Shield Fan Lobster]
Ibacus novemdentatus Gibbes, 1850 [Smooth Fan Lobster]
Parribacus antarcticus (Lund, 1793) [Sculptured Mitten Lobster]
Subfamily Theninae [Flathead Lobsters]
†***Thenus orientalis*** (Lund, 1793) [Moreton Bay Flathead Lobster]
Subfamily Scyllarinae [Locust Lobsters]
†***Scyllarus arctus*** (Linnaeus, 1858) [Small European Locust Lobster]

Scyllarus batei Holthuis, 1946 [Soft Locust Lobster]
Scyllarus caparti Holthuis, 1952
Scyllarus posteli Forest, 1963
Scyllarus pygmaeus (Bate, 1888) [Pygmy Locust Lobster]
INFRAORDER THALASSINIDEA [Mud Lobsters]
Superfamily Thalassinoidea
Family Callianassidae
Callianassa tyrrhena (Petagna, 1792) [Ghost Mud Lobster]
Family Upogebiidae
Upogebia pusilla (Petagna, 1792) [Mediterranean Mud Lobster]
INFRAORDER ANOMURA
Superfamily Paguroidea [Hermit Crabs]
Family Diogenidae
Clibanarius erythropus (Latreille, 1818) [Latreille's Hermit Crab]
Diogenes pugilator (Roux, 1829) [Roux's Hermit Crab]
Family Paguridae
Pagurus bernhardus (Linnaeus, 1758) [Bernard's Hermit Crab]
Pagurus prideaux Leach, 1815 [Prideaux's Hermit Crab]
Family Lithodidae [Stone Crabs]
Lithodes maja (Linnaeus, 1758) [Northern Stone Crab]
†*Lithodes santolla* (Molina, 1782) [Patagonian Royal Stone Crab]
Neolithodes grimaldii (A. Milne Edwards & Bouvier, 1894) [Grimaldi's Stone Crab]
Paralithodes brevipes (Brandt, 1850) [Brown King Stone Crab]
†***Paralithodes camtschaticus*** (Tilesius, 1815) [Alaskan King Stone Crab]
Paralithodes platypus (Brandt, 1850) [Blue King Stone Crab]
Paralomis bouvieri Hansen, 1908
Paralomis microps Filhol, 1884
Paralomis spectabilis Hansen, 1908
Superfamily Galatheoidea
Family Galatheidae [Squat Lobsters]
†***Cervimunida johni*** Porter, 1903 [Yellow Squat Lobster]
Galathea bolivari Zariquiey, 1950
Galathea intermedia Lilljeborg, 1851
Galathea squamifera Leach, 1815 [Leach's Squat Lobster]
Galathea strigosa (Linnaeus, 1767) [Strigose Squat Lobster]
Munida intermedia A. Milne Edwards & Bouvier, 1899
Munida rugosa (Fabricius, 1775) [Rugose Squat Lobster]
Munida sarsi Huus, 1935

Munida tenuimana G. O. Sars, 1872
†***Pleuroncodes monodon*** (H. Milne Edwards, 1837) [Red Squat Lobster]
Pleuroncodes planipes Stimpson 1858
Family Porcellanidae [Porcelain crabs]
Porcellana platycheles (Pennant, 1777) [Broad-clawed Porcelain Crab]
Pisidia bluteli (Risso, 1816)
Pisidia longicornis (Linnaeus, 1767) [Long-clawed Porcelain Crab]
Pisidia longimana (Risso, 1816)
INFRAORDER BRACHYURA [True Crabs]
Section Dromiacea
Superfamily Dromioidea
Family Dromiidae [Sponge Crabs]
Dromia personata (Linnaeus, 1758) [Linnaeus's Sponge Crab]
Section Archaeobrachyura
Superfamily Homoloidea
Family Homolidae [Box Crabs]
Homola barbata (Fabricius, 1793) [Hairy Box Crab]
Paramola cuvieri (Risso, 1816) [Risso's Box Crab]
Section Oxystomata [Narrow-mouth Crabs]
Superfamily Dorippoidea
Family Dorippidae [Demon-faced Crabs]
Medorippe lanata (Linnaeus, 1767) [Demon-faced Porter Crab]
Superfamily Leucosioidea
Family Leucosiidae [Nut Crabs]
Ebalia tuberosa (Pennant, 1777) [Pennant's Nut Crab]
Leucosia signata Paulson, 1875
**Myra fugax* (Fabricius, 1788)
Section Cancridae
Family Corystidae [Helmet Crabs]
Corystes cassivelaunus (Pennant, 1777) [Helmet Crab]
Family Atelecyclidae [Circular Crabs]
Atelecyclus rotundatus (Olivi, 1792) [Circular Crab]
Atelecyclus undecimdentatus (Herbst, 1783) [Broad Circular Crab]
Family Cancridae [Edible Crabs]
Cancer bellianus Johnson, 1861 [Madeiran Crab]
†***Cancer pagurus*** (Linnaeus, 1758) [European Edible Crab]
Section Brachyrhyncha
Superfamily Portunoidea
Family Geryonidae [Geryonid Crabs]

Chaceon affinis (A.Milne Edwards & Bouvier, 1894) [Red Geryonid Crab]
†***Chaceon maritae*** Manning & Holthuis, 1981 [West African Geryonid Crab]
Family Portunidae
Subfamily Polybiinae [Swimming Crabs]
Liocarcinus depurator (Linnaeus, 1758) [Harbour Swimming Crab]
Liocarcinus holsatus Fabricius, 1798 [Flying Swimming Crab]
†***Necora puber*** (Linnaeus, 1767) [Velvet Swimming Crab]
Polybius henslowii Leach, 1820 [Sardine Swimming Crab]
Subfamily Portuninae
*†***Callinectes sapidus*** Rathbun, 1896 [Chesapeake Bay Swimming Crab]
**Charybdis helleri* (A.Milne Edwards, 1867)
**Charybdis longicollis* Leene, 1938
**Charybdis merguiensis* de Man, 1887
**Charybdis sexdentatus* (Herbst, 1783)
†***Portunus pelagicus*** (Linnaeus, 1758) [Blue Swimming Crab]
†***Scylla serrata*** (Forskal, 1775) [Mud Swimming Crab]
**Thalamita poissonii* (Audouin, 1826)
Subfamily Carcininae [Shore Crabs]
†*Carcinus aestuari* Nardo,1847 [Mediterranean Shore Crab]
Carcinus maenas (Linnaeus, 1758) [North Atlantic Shore Crab]
Superfamily Xanthoidea
Family Goneplacidae [Goneplacid or Angular Crabs]
**Eucrate crenata* (de Haan, 1835)
Family Xanthidae [Xanthid crabs]
Subfamily Xanthinae
**Dyspanopeus sayi* (Smith, 1869) [Caribbean Xanthid Crab]
Eriphia verrucosa (Forskal, 1775) [Warty Xanthid Crab]
**Rhithropanopeus harrisii* (Gould, 1841) [Dwarf Xanthid Crab]
Xantho incisus Leach, 1814 [Furrowed Xanthid Crab]
Xantho incisus granulicarpus Forest, 1953
Xantho pilipes A. Milne Edwards, 1867 [Less Furrowed Crab]
Subfamily Pilumninae
**Heteropanope laevis* Dana, 1852
**Pilumnoides inglei* Guinot & Macpherson, 1987
**Pilumnopeus vauquelini* (Savigny & Audouin, 1819)
Pilumnus hirtellus (Linnaeus, 1761) [Bristly Xanthid Crab]
Superfamily Pinnotheroidea
Family Pinnotheridae
Subfamily Pinnotherinae [Pea Crabs]

Pinnotheres pisum (Linnaeus, 1761) [Linnaeus's Pea Crab]
Pinnotheres pisum pectunculi (Hesse, 1861)
Nepinnotheres pinnotheres (Linnaeus, 1758) [Pinnae Pea Cab]
Superfamily Grapsioidea
Family Grapsidae
Subfamily Grapsinae [Rock Crabs]
**Brachynotus sexdentatus* (Risso, 1826) [Mediterranean Rock Crab]
Pachygrapsus marmoratus (Fabricius, 1787) [Marbled Rock Crab]
Subfamily Varuninae [Varunid Crabs]
*†***Eriocheir sinensis*** de Haan, 1835 [Mitten Crab]
Eriocheir japonica De Haan, 1835 [Japanese Mitten Crab]
Superfamily Ocypodidea
Family Ocypodidae [Fiddler and Running Crabs]
†***Uca tangeri*** (Eydoux, 1835) [Moroccan Fiddler Crab]
Superfamily Potamoidea
Family Potamidae [Freshwater Crabs]
Subfamily Potaminae
Potamon fluviatile algeriense Bott, 1967
Potamon fluviatile fluviatile (Herbst, 1785) [Italian Freshwater Crab]
Potamon ibericum (Bieberstein, 1809) [Bieberstein's Freshwater Crab]
Potamon potamios (Olivier, 1804) [Aegean Freshwater Crab]
Section Oxyrhyncha
Superfamily Majoidea
Family Majidae [Spider Crabs]
Subfamily Inachinae
Inachus dorsettensis (Pennant, 1777) [Scorpion Spider Crab]
Macropodia rostrata (Linnaeus, 1761) [Long-legged Spider Crab]
Subfamily Majinae
Maja crispata Risso, 1827 [Lesser Spider Crab]
Maja goltziana d'Oliveira, 1888
†***Maja squinado*** (Herbst, 1788) [Thorn-back Spider crab]
†*Chionoecetes bairdi* Rathbun, 1893 [Snow Queen Spider Crab]
†*Chionoecetes opilio elongatus* Rathbun, 1924 [Pacific Snow Spider Crab]
†***Chionoecetes opilio opilio*** (Fabricius, 1788) [Atlantic Snow Spider Crab]
†*Chionoecetes tanneri* (Rathbun, 1893) [Tanner Spider Crab]
Subfamily Pisinae
Pisa armata (Latreille, 1803) [Gibb's Spider Crab]

FURTHER READING

Anon., (1985) *International Code of Zoological Nomenclature adopted by the XX General Assembly of the International Union of Biological Sciences.* Third Edition. London. International Trust for Zoological Nomenclature in Association with the British Museum (Natural History).

Bowman, T.E. and Abele, L.G. (1982) Classification of the Recent Crustacea, The Systematics, the Fossil Record and Biogeography, in *Volume 1, The Biology of Crustacea* (ed. L.G. Abele) Academic Press, New York.

Calman, W.T. (1949) *The Classification of Animals. An Introduction to Zoological Taxonomy.* Methuen & Co, London.

Cherry, G.N. (1992) *Some of the Shellfish Traded and Consumed in Europe: Together with an Interpretation of their Scientific Names.* Eastbourne, Sussex, privately published.

Gotto, H. (1982) *Animal Taxonomy.* Studies in Biology **143**. Institute of Biology. Edward Arnold, London.

Jaeger, E.C. (1950) *A Source-Book of Biological Names and Terms.* Charles C. Thomas, Springfield, Illinois.

Lincoln, R.J., Boxshall, G.A., Clark, P.F. (1982) *A Dictionary of Ecology, Evolution and Systematics.* Cambridge University Press, London, New York, etc.

Noel, P.Y. (1992) Cle preliminaire d'identification des Crustacea Decapoda de France et des principales autres espèces d'Europe. *Collection Patrimoines Naturels* **9**. *Série Patrimoine Scientifique.* Museum National d'Histoire Naturelle, Paris.

Williams, A.B., *et al.* (1989) Common and scientific names of aquatic invertebrates from the United States and Canada. Decapod Crustacea. *American Fisheries Society Special Publication* **17**. Bethesda, Maryland.

3

Overview of groups

The species of decapods listed on pp 14-19 can be superficially classified into lobster-like and crab-like forms represented by: (a) crayfishes and lobsters; (b) crawfishes and shovel-nosed lobsters; (c) mud lobsters; (d) false lobsters, false crabs and hermit crabs; (e) true crabs.

3.1 FRESHWATER CRAYFISHES AND CLAWED LOBSTERS

Both freshwater crayfishes and marine clawed lobsters (infraorder Astacidea) comprises in excess of 750 species inhabiting bottom substrates of rivers, lakes and seas (benthic habitats). They are distributed throughout most regions of the world, but freshwater crayfishes do not occur naturally in Africa and parts of Asia, although at least one species has been introduced into these two regions by Man.

The crayfishes and marine clawed lobsters describe in this book share a number of obvious features. Both have a substantially hard and usually relatively smooth carapace (**cr**, fig. 3.1, A) and each first (**pr1**) to third (**pr3**) pairs of pereiopods terminate in a 'pincer or claw'=chela (**ce**); the first pair are large and powerful, the second and third pairs are much smaller. The telson (**tl**) and uropods (**ur**) form a broad 'tail fan'.

The carapace of freshwater crayfishes is without a medial groove or spinous ridge and the telson has a partly or fully developed transverse suture (fig. 3.1, B). Males of northern hemisphere crayfishes (superfamily Astacoidea) have highly modified first and second pairs of gonopods (fig. 4.4, C), the first pair (**pl1**) are formed as an almost complete tube for conducting sperm.

The carapace of marine clawed lobsters (superfamily Nephropoidea) has a medial groove (**mv**, fig. 3.1, A) or spinosed ridge and there is no transverse suture on the telson. The tube-like structure of the male's first gonopod is incomplete and the second is hardly modified (fig. 1.1, C).

3.2 CRAWFISHES AND SHOVEL-NOSED LOBSTERS

Crawfishes (fig. 3.1, C) and the related Shovel-nosed Lobsters (fig. 3.1, D) all belong to the superfamily Palinuroidea of the infraorder Palinura. There are probably in excess of 120 species and all live in benthic habitats throughout tropical to temperate oceans of the world. These lobsters do not have the first (**pr1**) to third (**pr3**) pairs of pereiopods terminating as chelae and the first pair are not noticeably stout. The posterior part of the 'tail fan' is membranous and flexible.

3.3 MUD LOBSTERS

There are about 350 known species of mud lobsters (infraorder Thalassinidea). They can be recognised by the relatively thin and pliable carapace that has a conspicuous longitudinal suture (**ls,** fig. 3.1, E) on each side. The abdomen is very flexible and the segments usually loosely articulate.

3.4 FALSE LOBSTERS AND CRABS

A large number (probably over 1,600 species) of lobster and crab-shaped forms, sometimes referred to as false lobsters and false crabs, belong to the infraorder Anomura. The name anomura means 'irregular tailed' [*anomos* Gr: irregular] but this feature is applicable only to hermit crabs (fig. 3.2, A) and Stone Crabs (B), both of which belong to the superfamily Paguroidea. The abdomen of marine hermit crabs is concealed within an empty gastropod shell.

False lobsters, false crabs and hermit crabs are distinguished from the true crabs (section 3.5) by having the second pair of antennae (**an2,** fig. 3.2, B-D) placed to the outside of the eyes. The sides of the carapace are thinly calcified or formed of thin plates. The fifth pair of pereiopods (**pr5,** B, C) are small, thin and (hermit crabs excepted) usually tucked beneath the posterior margins of the carapace but are shown exposed in the illustrations. Thus only four pairs of limbs are usually visible compared with the five conspicuous pairs of clawed lobsters (section 3.1), crawfishes (3.2), and the true crabs (3.5).

Anomuran squat lobsters (fig. 3.2, C) and porcelain crabs (D) (superfamily Galatheoidea) have symmetrical abdomens usually loosely pressed against the cephalothorax but shown extended here for the squat lobster illustrated.

3.5 TRUE CRABS

There are in excess of 4,500 species of true crabs (fig. 3.2, E) that all belong to the infraorder Brachyura. An overview of the various sections and families are appraised in section 10.6.

In comparison to false crabs the second pair of antennae (**an2**) are placed well to the inside of the eyes and the flagella of these appendages are very short. The fifth pair of pereiopods (**pr5**) are conspicuous and always visible.

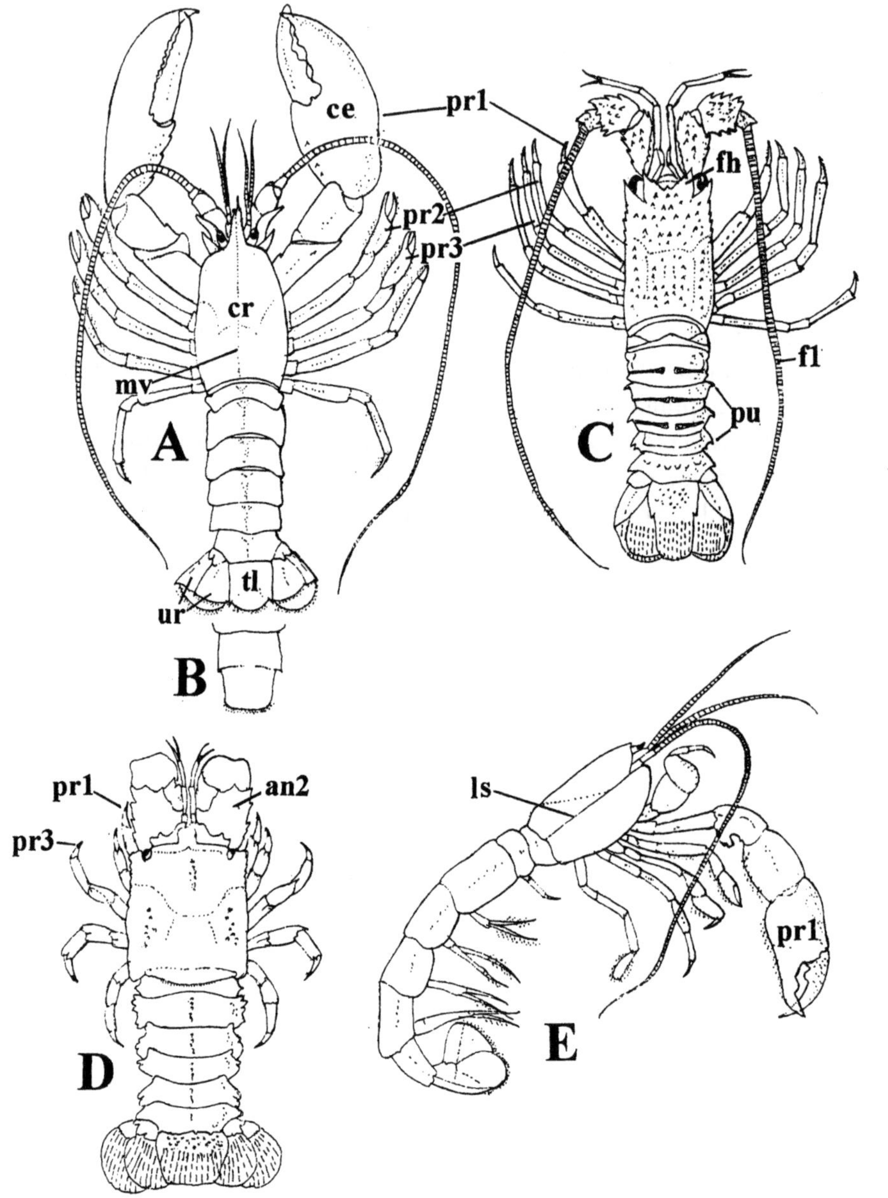

Fig. 3.1 A. Marine Clawed Lobster. B. Telson of freshwater crayfish. C. Spiny crawfish. D. Shovel-nosed lobster. E. Mud lobster. For explanation of annotations see text.

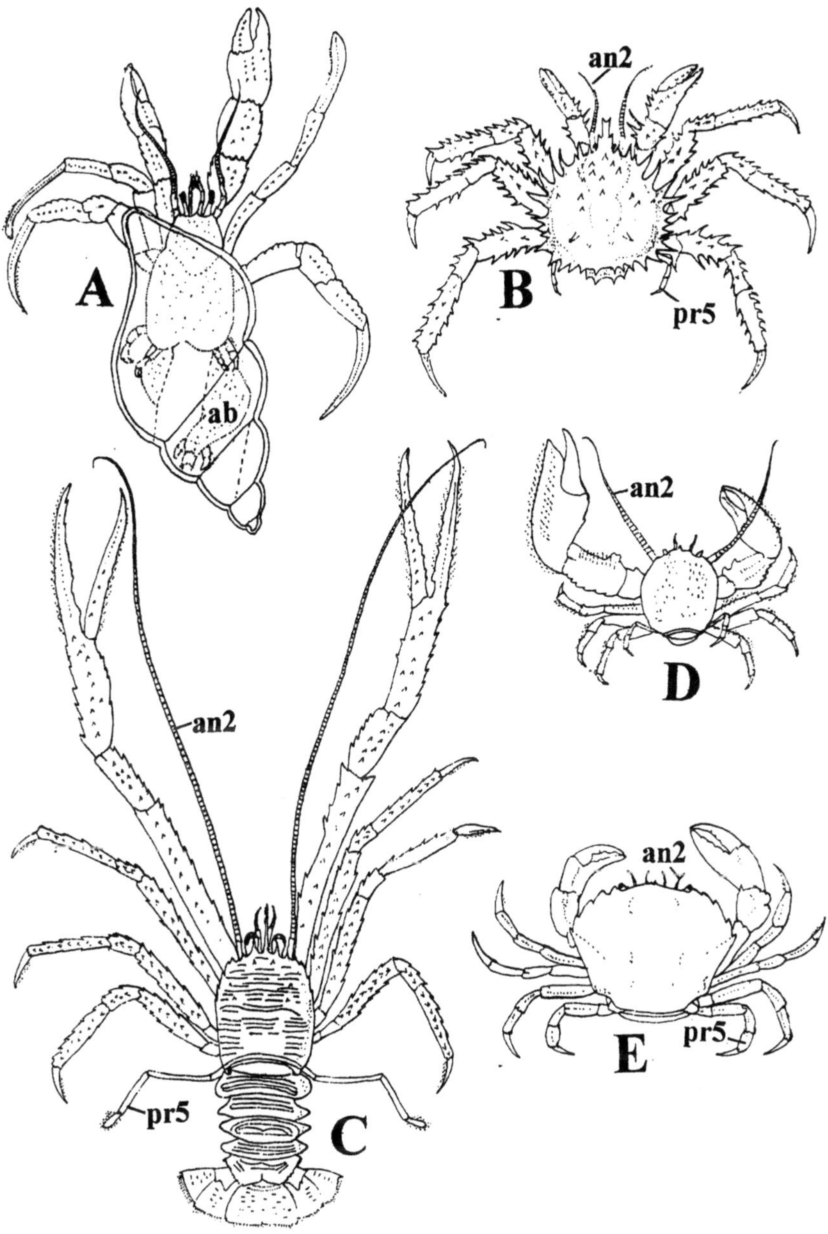

Fig. 3.2 A. Hermit crab. B. Stone crab. C. Squat lobster. D. Porcelain crab. E. True crab. For explanation of annotations see text.

FURTHER READING

Allen, J.A. (1967) *The Fauna of the Clyde Sea Area. Crustacea: Euphausiacea and Decapoda.* Scottish Marine Biological Association, Millport.

Falciai, L and Minervini, R. (1992) *Guida dei crostacei decapodi d'Europa.* Franco Muzzio Editore, Padova.

Fischer, W (ed). (1978) *Volume VI, Lobsters, Shrimps, Prawns, True Crabs,* in FAO Species Identification Sheets, Western Central Atlantic. FAO, Rome.

Fischer, W., Bianchi, G., Scott, W.B. (eds). (1981) *Volume V, Lobsters, Volume VI, Shrimps and Prawns, True Crabs,* in FAO Species Identification Sheets, Eastern Central Atlantic. FAO, Rome.

Ingle, R.W. (1995) *The UFAW Handbook on the Care & Management of Decapod Crustaceans in Captivity.* Universities Federation for Animal Welfare, Herts, England, (pages 24-35 for illustrated overview of decapods).

McLaughlin, P.A. (1980) *Comparative Morphology of Recent Crustacea.* Freeman & Co, San Francisco.

Noel, P.Y. (1992) Cle preliminaire d'identification des Crustacea Decapoda de France et des principales autres espèces d'Europe. *Collection Patrimoines Naturels* **9**. *Série Patrimoine Scientifique.* Museum National d'Histoire Naturelle, Paris.

Schmitt, W.L. (1965) *Crustaceans.* Ann Arbor. The University of Michigan Press, USA & Ambassador Books Ltd, Toronto, Canada.

Zariquiey Alvarez, R. (1968) Crustáceòs Decápodos Ibéricos. *Investigación Pesquera,* **32**, 1-510.

4

Freshwater Crayfishes

(Astacoidea)

Crayfishes inhabiting the northern hemisphere and discussed below belong to the superfamily Astacoidea. Species of this superfamily have the first and second pairs of pleopods present. The first pair of the female may be quite small, but for the male these, and the second pair, are modified as sperm transference organs [gonopods, figs. 4.4-4.6, C]. These crayfishes also have a complete transverse suture on the telson (fig. 3.1, B) and the posterior part of the uropods (**ur**, fig. 1.1, A) are well calcified.
These combined features distinguish astacid crayfishes from the parastacid species (see section 11.1) imported from the southern hemisphere.

4.1 NATIVE SPECIES

Expert opinion differs as to the precise number of crayfish species and subspecies inhabiting Europe (see section 4.5). Six are recognised here. The Noble Crayfish (*Astacus astacus astacus*), two subspecies of Long-clawed Crayfish, (*Astacus leptodactylus*), the Caspian Crayfish, (*Astacus pachypus*), the Stone Crayfish, (*Austropotamobius torrentium*) and the White-clawed Crayfish, (*Austropotamobius pallipes*). The last mentioned is the only species naturally occurring in the UK and there is at least one known occurrence of Noble Crayfishes living wild in the British Isles. By comparison, the Long-clawed Crayfish is now known from several localities in southern England.

4.2 IMPORTED SPECIES

At least three crayfish species have been imported from America and two from Australia for aquaculture, human consumption or for sale to aquarists. Descriptive accounts of these are given in chapter 11.

The importance of these crustaceans in crayfish farming is discussed in section 4.4.

4.3 LIFE HISTORY

The typical life history is described for the White-clawed Crayfish. Mating occurs between September and November. Females as small as 4 cm total length are able to mate successfully. Freshwater crayfishes and marine clawed lobsters (section 5) mate in a similar manner. The male grasps the female and turns her ventral side uppermost before depositing spermatophores onto her sternum (fig. 4.1, A). The eggs are usually spawned two or three days after mating. Largest females of approximately 9 cm total length may spawn as many as 150 eggs.

The eggs are large and carried attached to special setae [oosetae] on the female's pleopods [swimmerets] where they are incubated for seven to eight months and regularly preened by the brooding female; preening is carried out by the posterior pairs of pereiopods.

The hatchlings burst from the egg capsules in an advanced state of development and remain attached to the female's pleopods from three to seven days during which time they feed on their reserves of egg yolk (fig. 4.1, B).

They then moult to a form similar to the adults and leave the parent female to forage, but will return to reattach to the pleopods if danger threatens. Following a second moult, some ten days later, the young hatchlings now measure 13-15 mm in length and adopt a totally independent life style.

4.4 CRAYFISH FARMING

4.4.1 Review of species

In many European countries freshwater crayfishes are an important delicacy and command high prices in fish markets and restaurants. Although some

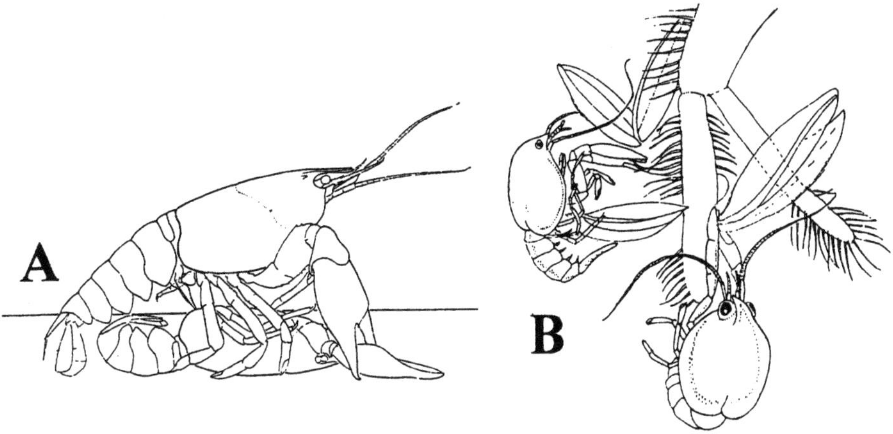

Fig. 4.1 A. Freshwater crayfish, final stage of mating-male depositing spermatophores on female's mid-sternum. B. Almost fully developed hatchling crayfishes 8.5-9 mm in length, attached to swimmeret setae of female.

consumer demands are met by locally fished crayfishes there is now an ever increasing aquaculture industry for these crustaceans currently limited to a few species.

The most important of these at present is the Signal Crayfish (see section 11.1) large numbers of which have been imported since the 1960's into European countries. The original intention was to use these to supplement or replace endemic crayfish stocks that had been decimated by crayfish plague (4.4.3). The focus of these primary introductions occurred in Sweden, a country where crayfishes are held in high esteem as a food item. The production rates of the successfully established populations however, fell far short of market demands and led to studies for intensive culture methods of Signal Crayfishes. This work was undertaken by the organisation AB Simontorps Akvelslaboratorium at Blentarp, Sweden, and in 1970 the first cultured juveniles were introduced into natural water bodies containing naturalised stocks and were also offered for sale to prospective crayfish farmers.

Juveniles of Signals introduced into a farm in Dorset, England, produced by the Simontorps Laboratory, and the subsequent franchise obtained from that supplier, formed the bases of crayfish farming in the UK. The growth of this industry led to the establishment of the British Crayfish Marketing Association in 1982 (now disbanded) to assist crayfish farmers to market good quality crayfishes and maintain prices; there is an estimated UK market of approximately 2000 tonnes.

The Noble Crayfish (p. 36) has always been considered a highly esteemed food item and was farmed for many centuries in parts of Europe. Western Russia, Finland and Sweden exported considerable quantities. However, by about 1960 production of this species had almost entirely ceased due to progressive decimation of stocks by crayfish plague. The species is now only experimentally cultured.

Because of its large size the Long-clawed Crayfish (p. 38) has always been commercially fished and its economic importance increased in proportion to the decline of the Noble Crayfish. Much of the production of Long-clawed Crayfishes originates in the USSR and Turkey and until recently it was an important part of the Turkish economy. Several countries, particularly France, Germany and Spain are rearing the species on an experimental commercial basis.

The White-Clawed Crayfish (p. 34) has never been considered particularly suitable for commercial farming because of its small size at maturity, but moderate quantities have always been fished and consumed locally although it is now a protected species in the UK. In Spain the species is currently farmed.

The North American Striped Crayfish, (see section 11.1) has never been considered a significantly marketable species because it is small and the narrow claws contain little meat. Nevertheless, its high productivity and toleration of eutrophic conditions and slight pollution make it very suitable for pond culture.

The Red Swamp Crayfish (section 11.1) is extensively farmed in the USA and has been imported into many countries for commercial culture. Its introduction into Africa resulted in the build up of large naturalised populations in Kenya from where considerable quantities have been exported to European markets. In southern Spain a successful fishery has been developed but feral populations cause considerable damage to rice crops. Frozen whole and boiled, these crayfishes are now gaining a more prominent place in European markets and replacing dwindling imports of Long-clawed Crayfishes, stock of which have become decimated through crayfish plague.

4.4.2 Farming and culture

Two methods of farming crayfishes are currently practiced. The first is by stocking natural or artificial ponds that contain adequate aquatic vegetable and animal life to support growth of introduced juvenile crayfishes. The population is allowed to become established and then periodically fished by netting or trapping. Crayfishes of marketable sizes are retained and the remainder returned to the ponds.

For the second method (semi-intensive farming) juvenile crayfishes are introduced into growing-on ponds (fig. 4.2). These specially dug canal-like parallel enclosures each has a controlled inlet and outlet. The development of natural vegetation in these ponds is encouraged but artificial food is often supplied (eg. trout pellets). Crayfishes are harvested by trapping in suitably baited funnel-shaped traps (fig. 4.3, A), by netting or draining the ponds.

Many crayfish farmers now produce their own juvenile crayfishes for stocking growing-on ponds. A method often used involves the selection of males and females that are held in ponds shortly before the breeding season. After mating the females are removed to brood-stock ponds and overwintered. Shortly before the eggs are due to hatch the females are placed into hatching tanks on perforated grids of a sufficiently large enough diameter to allow second stage hatchlings to fall through to the bottom of the tank from where they are collected after removal of the females. The hatchlings are usually reared in nursery tanks until they reach a suitable size for stocking growing-on ponds.

An alternative but still somewhat experimental method for commercial rearing of juvenile crayfishes is achieved by removing all the eggs from the female and incubating these in specially constructed vessels (fig. 4.3, B) similar to those used for hatching fish eggs.

4.4.3 Crayfish plague

This is a disease of crayfish caused by infection of the body tissues with the fungal pathogen *Aphanomycetes astaci* Schikora, 1903. Infections usually result in high mortality rates among populations of native crayfishes and the fungus is easily transmitted by contaminated fishing nets and by transporting unknown infected crayfishes and water to and from farms. The extensively farmed Signal Crayfish is resistant to the fungus and can act as a carrier.

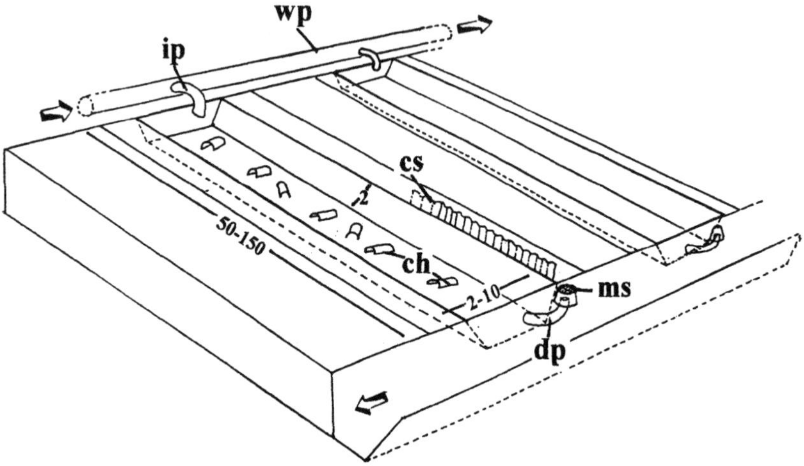

Fig. 4.2 Growing-on ponds designed for semi-intensive farming of crayfishes. These are often netted to prevent escapes or predation: **ch** 'hides' for crayfishes; **cs** corrugated plastic sheets 20-25 cm high inserted around pond margin to prevent escapes; **dp** swing-arm outflow pipe discharging water into gully for return to source; **ip** swing-arm inflow pipe above water level; **ms** mesh covering outflow pipe to prevent escapes; **wp** pipe carrying inflowing water from stream or river; arrows indicate direction of water flow; measurements in metres.

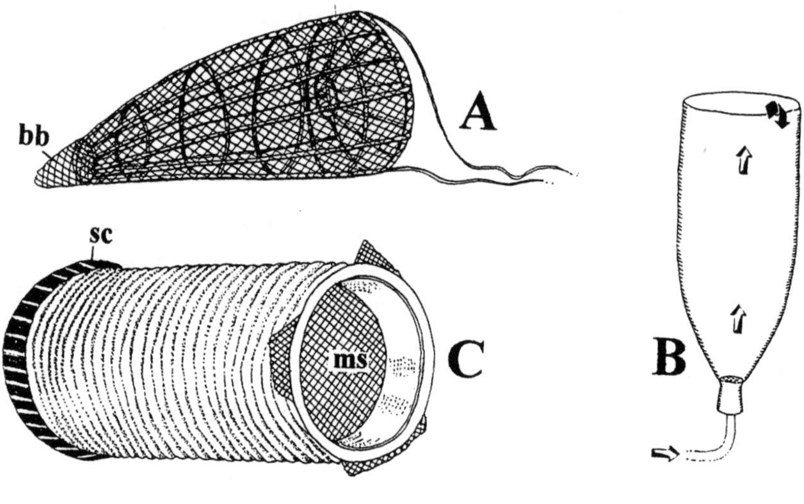

Fig. 4.3 A. Crayfish trap used for fishing growing-on ponds. B. Crayfish egg incubator-water overflows into trough beneath jar. C. Container used for transporting juvenile crayfishes for seeding ponds: **bb** bait bag; **ms** mesh; **sc** screw cap; arrows indicate direction of water flow.

4.5 IDENTIFICATION

There are various alternative views about generic and specific status of European crayfishes. The genera *Astacus* and *Austropotamobius* probably should be maintained; species of both do not interbreed naturally.

Species belonging to the genus *Astacus* have two post-orbital (**pd**, figs 4.5, 4.6, A) ridges; only one pair are present in *Austropotamobius* (fig. 4.4, A).

In recent studies three subspecies of *Astacus astacus* have been recognised based upon slight morphological differences: *A. astacus astacus*-The Noble Crayfish, *A. astacus balcanicus* and *A. astacus colchicus*.

At least five subspecies of *Astacus leptodactylus* have been described: *A. leptodactylus leptodactylus* and *A. leptodactylus salinus*-(both cited here as the Long-clawed Crayfish), *A. leptodactylus eichwaldi*, *A. leptodactylus. cubanicus*, and *A. leptodactylus caspicus.* Except for the first two mentioned, these subspecies are separated on indistinct and, in the author's opinion, variable features. However, as the name implies the two subspecies of Long-clawed Crayfishes can be readily recognised by the long thin chelipeds. These are straight in *A. leptodactylus leptodactylus* (fig. 4.6, A) and distinctly curved in *A. leptodactylus salinus* (fig. 4.6, B). The long rostrum separate both these subspecies from *A. astacus astacus* the Noble Crayfish that has a short rostrum and also broad chelipeds (fig. 4.5, A).

Astacus pachypus (Caspian Crayfish) is confined to the Caspian and Azov regions of Russia. This species differs from *A. astacus astacus* by having apical spines on the posterior pair of carapace post-orbital ridges, a dorso-basal protuberance on the endopod of the male's second gonopod and the involuted part extending to half the endopod length.

Austropotamobius pallipes (White-clawed Crayfish) is distinguished from the more rare European *A. torrentium* (Stone Crayfish) by having denticles just behind the cervical groove (**cg**, fig. 4.4, A) and a low but conspicuous medial apical ridge on the rostrum (both features are absent in the Stone Crayfish). The lateral teeth of the rostrum are also well developed, (they are obtuse in the Stone Crayfish).

The four subspecies of *A. pallipes* (*A .pallipes carsicus*, *A. pallipes fulcisianus*, *A. pallipes italicus, A. pallipes lusitanicus*) and the species *A. (Atlantoastacus) bernhauseri* were established on very indistinct and variable features; they are generally not accepted as valid.

Austropotamobius pallipes (Lereboullet, 1858)
Fig. 4.4, Pl.1.
COMMON SYNONYMS: *Astacus pallipes*; *Austropotamobius (Atlantoastacus) pallipes pallipes*; *Austropotamobius (Atlantoastacus) pallipes italicus*.
DERIVATION: austro-denoting presence in Austria; potamos (Gr)-a river; *pallipes* derived from pallidus (Lt)-alluding to white colour on undersides of pereiopods.
VERNACULAR NAMES: White-clawed Crayfish; White-footed Crayfish (UK); Flodkräfta (Sw); Dohlenkrebs (Gn); Rivierkreeft (Nd); Écrevisse à Pieds Blancs (Fr); Cangrego de Rio (Sp); Lagostim do Rio (Pt); Gambero di Fiume (It); Karavida tou Glykou Nerou (Gr); Göl Istakozu (Tr).
DISTINCTIVE FEATURES: rostrum (**rs,** fig. A) converging distally, with small apical median ridge (**mr,** fig. B), lateral teeth (**lt**) acute, carapace with a single postrostral ridge (**pd**), small denticles behind cervical groove (**cg**). Male-distal lobes of first pleopod (**pl1,** fig. C) equal in size, exopod (**ex**) of second pleopod (**pl2**) reaching, at most, to base of endopod (**en**) flagellum (**fl**), involuted part (**it**) of endopod measuring less than about one third length of endopod, a protuberance (**pn**) on inner basal part of endopod.
COLOUR: variable, often greenish brown but ranging from almost black to bluish; ventral surfaces of limbs and underside of body orange to whitish.
TOTAL LENGTH: about 10 cm in UK, exceptionally 13.5 cm in Europe.
DISTRIBUTION: ranging from 56° N in Great Britain to 38° S in Europe (Spain); from 8° W (Ireland) to 16°E in Europe (Yugoslavia).

The White-clawed Crayfish occurs in freshwater streams, rivers and reservoirs, beneath rocks and stones, sometimes in shallow burrows.

GENERAL INFORMATION: This crayfish is not generally considered a viable species for aquaculture because of slow growth and small size. The age of large specimens in northern England were estimated at eleven years.

In the UK taking this species from the wild is now prohibited. In France fishing is allowed only during ten days each year at different periods with a minimum legal size of 9 cm. In Switzerland fishing is permitted for 12 weeks, minimum size 10 cm, and in Italy for 9 months in most regions at a legal size 7 cm, but prohibited in Liguria, Lombardia and Piemonte. In Spain fishing is allowed only during public holidays and on two days of each week from 21 June-31 August.

Flavour of the meat is considered rather strong when compared to that of the Noble Crayfish; in France this crayfish is often used for making stews.

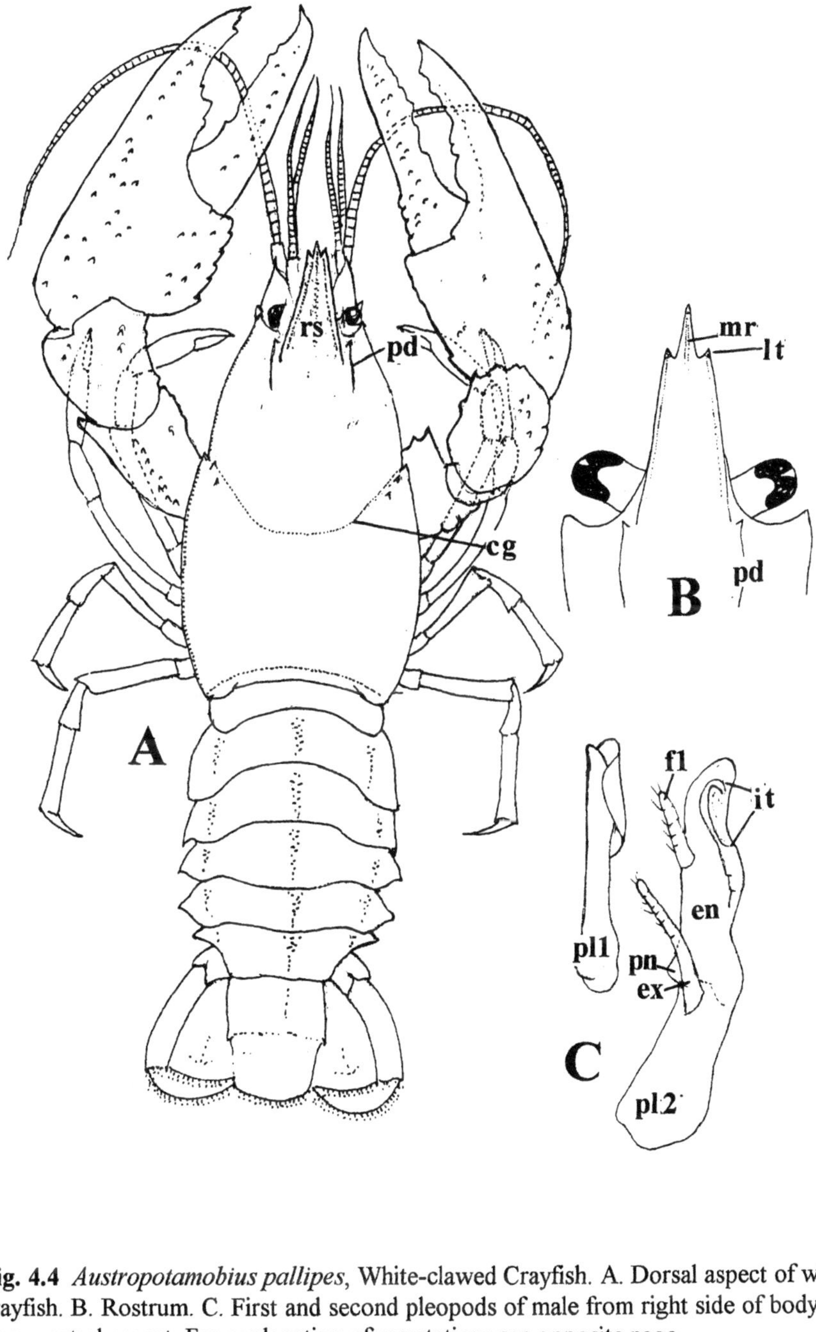

Fig. 4.4 *Austropotamobius pallipes*, White-clawed Crayfish. A. Dorsal aspect of whole crayfish. B. Rostrum. C. First and second pleopods of male from right side of body and from ventral aspect. For explanation of annotations see opposite page.

Astacus astacus astacus (Linnaeus,1758)
Fig. 4.5, Pl. 2.
COMMON SYNONYMS: *Astacus fluviatilis; Astacus nobilis; Potamobius fluviatilis; Astacus (Astacus) astacus.*
DERIVATION: astakos (Gr)-a kind of lobster or crayfish.
VERNACULAR NAMES: Noble Crayfish (UK); Ferskvannskreps (Nw); Edelkrebs (Gr); Écrevisse à Pieds Rouges (Fr); Gambaro (It); Tatli su Istakozu (Tr).
DISTINCTIVE FEATURES: rostrum (**rs,** fig. A) not distinctly converging distally, apex long, median ridge (**mr**, fig. B) usually granular; carapace with two distinct post-orbital ridges (**pd**), second pair often distally blunt; larger cheliped (**pr1**) usually with concavity on propodal (**pro**) inner margin. Male-distal lobes of first pleopod (**pl1,** fig. C) unequal, exopod (**ex**) of second pleopod (**pl2**) reaching beyond base of endopod (**en**) flagellum (**fl**), involuted part (**it**) of endopod measuring about one quarter length of endopod.
COLOUR: variable, black, olive brown, yellowish green, ruby red; extremities of limbs often reddish, sometimes mottled.
TOTAL LENGTH: about 10 cm, exceptionally 20 cm.
DISTRIBUTION: western Central Europe, Finland and Norway to approx. 61° N; eastward to eastern Lithuania, westward to north western France; southward to approx. 43° S. Also Spain and Portugal (but probably introduced). Introduced into Britain.

The Noble Crayfish inhabits freshwater rivers and lakes and occurs beneath rocks and stones, also burrows into suitable (ie. clay) substrates.
GENERAL INFORMATION: named Noble Crayfish because of alluring reddish colour and pleasant odour when being cooked. Always in demand in Europe and prized for culinary excellence.

Probably has maximum life span of about seven years.

In some Lithuanian lakes fishing is prohibited to allow overfished stocks to recover. In other lakes fishing is permitted only from 15 July-1 November with minimum size limit of 11 cm. In Poland, Sweden, Denmark and France minimum allowed capture size is 9.cm, in Norway 9.5 cm, in Hungary and in Finland 10 cm. In Poland only males are allowed to be fished from 16 March-14 October; in Norway fishing is permitted from 7 April (and in Hungary from 1 June) until October.

The species may have been introduced into the UK in previous centuries for farming and in this century for educational and research purposes. Some recent importations into the UK for experimental aquaculture have escaped from captivity and are thriving in a reservoir.

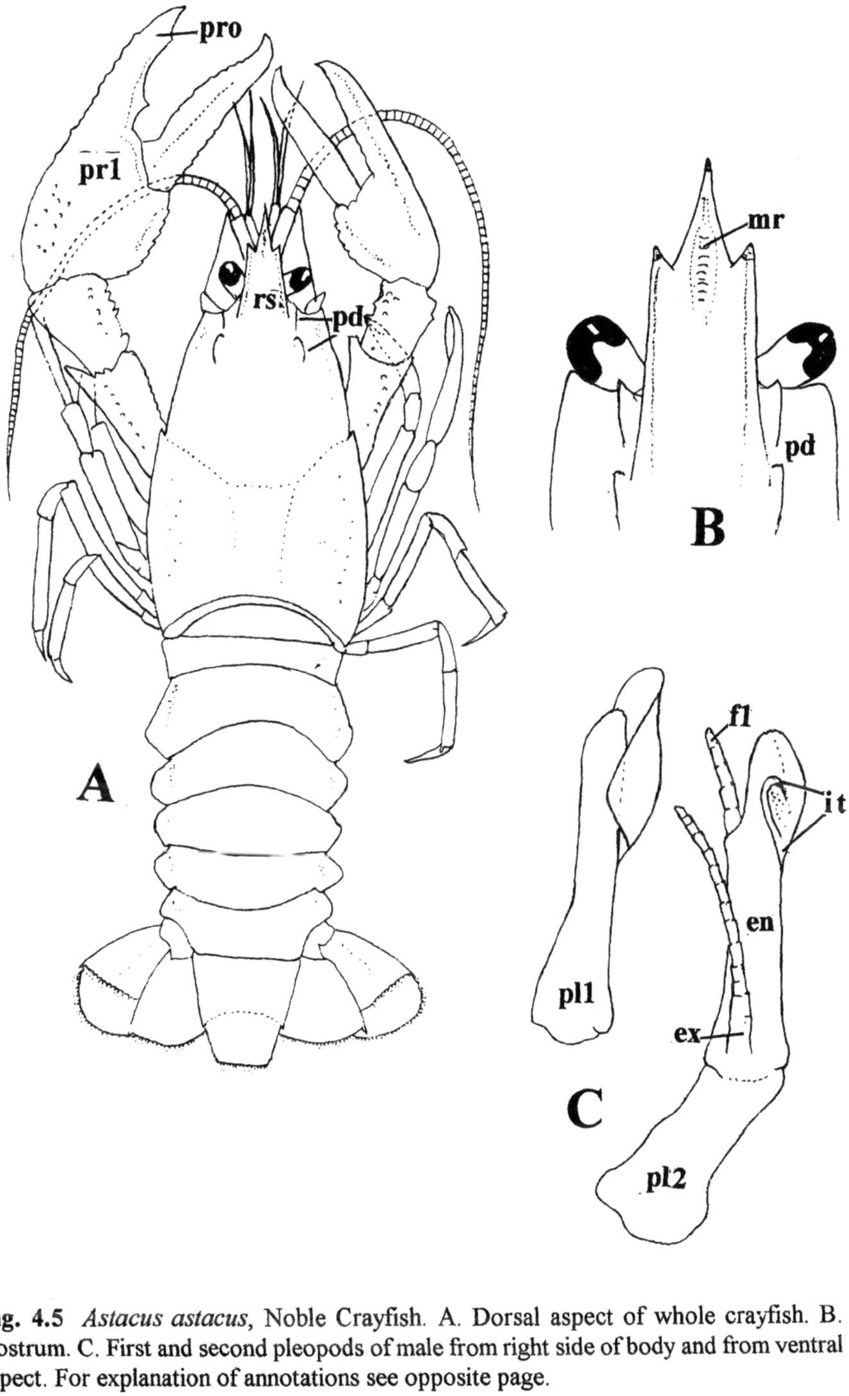

Fig. 4.5 *Astacus astacus,* Noble Crayfish. A. Dorsal aspect of whole crayfish. B. Rostrum. C. First and second pleopods of male from right side of body and from ventral aspect. For explanation of annotations see opposite page.

Astacus leptodactylus leptodactylus (Eschscholtz, 1823)
Fig. 4.6, Pl.3.
COMMON SYNONYMS: none.
DERIVATION: astakos (Gr)-a kind of lobster or crayfish; leptodes (Gr)-thinness; daktylos (Gr)-a finger, alluding to the thin chelipeds.
VERNACULAR NAMES: Long-clawed Crayfish, Turkish, Galician, Swamp, Marsh or Pond Crayfish (UK); Sumpfkrebs (Gn); Écrevisse à Pattes grêles (Fr); Gambero di Fiume Turco (It); Kerevit (Tn).
DISTINCTIVE FEATURES: rostrum (**rs**, fig. A) long, slightly converging distally and somewhat narrow, carapace with two distinct post-orbital ridges (**pd**), second pair with obvious distal spine; chelipeds (**pr1**, figs A, B) noticeably long and narrow, without concavity on propodal (**pro**) inner margin. Male-distal lobes of first pleopod (**pl1**, fig. C) unequal, exopod (**ex**) of second pleopod (**pl2**) reaching beyond base of endopod (**en**) flagellum (**fl**), involuted part of endopod (**it**) measuring slightly less than half length of endopod, a protuberance (**pn**) on inner basal part of endopod.
COLOUR: variable, olive green, yellowish brown, sometimes tinged with red, occasionally black; chelipeds sometimes with brown spots.
TOTAL LENGTH: about 12-13 cm.
DISTRIBUTION: *A. leptodactylus leptodactylus* occurs in lower reaches of the River Don, in rivers Dniester, Volga and tributaries flowing into the Black, Caspian and Azov Seas. It also inhabits Crimean and Lithuanian rivers flowing into the Black Sea. It has been introduced into France, Eastern Germany and the British Isles.

Both subspecies occur in diverse habitats, deep and shallow lakes, ponds, streams and rivers, on all substrates.

GENERAL INFORMATION: *A. leptodactylus salinus* is distinguished from *A. leptodactylus leptodactylus* by having curved (sickle-shaped) chelipeds, the left is shown in fig. 4.6, B.

The subspecies *salinus* inhabits lakes and rivers of the Danube basin and extends into brackish water. In the Ukraine it is distributed eastward and occurs with the subspecies *leptodactylus* in regions of transition between steppes and forest. *A. leptodactylus leptodactylus* is thus known as the 'steppes form' and *A. leptodactylus salinus* as the 'forest form'. Both subspecies may be present in rivers of central France and eastern Germany. *A. leptodactylus leptodactylus* now inhabits many ponds and canals in southern Britain.

The Long-clawed Crayfish is exported from Russia and Turkey and is now experimentally cultured in parts of Europe.

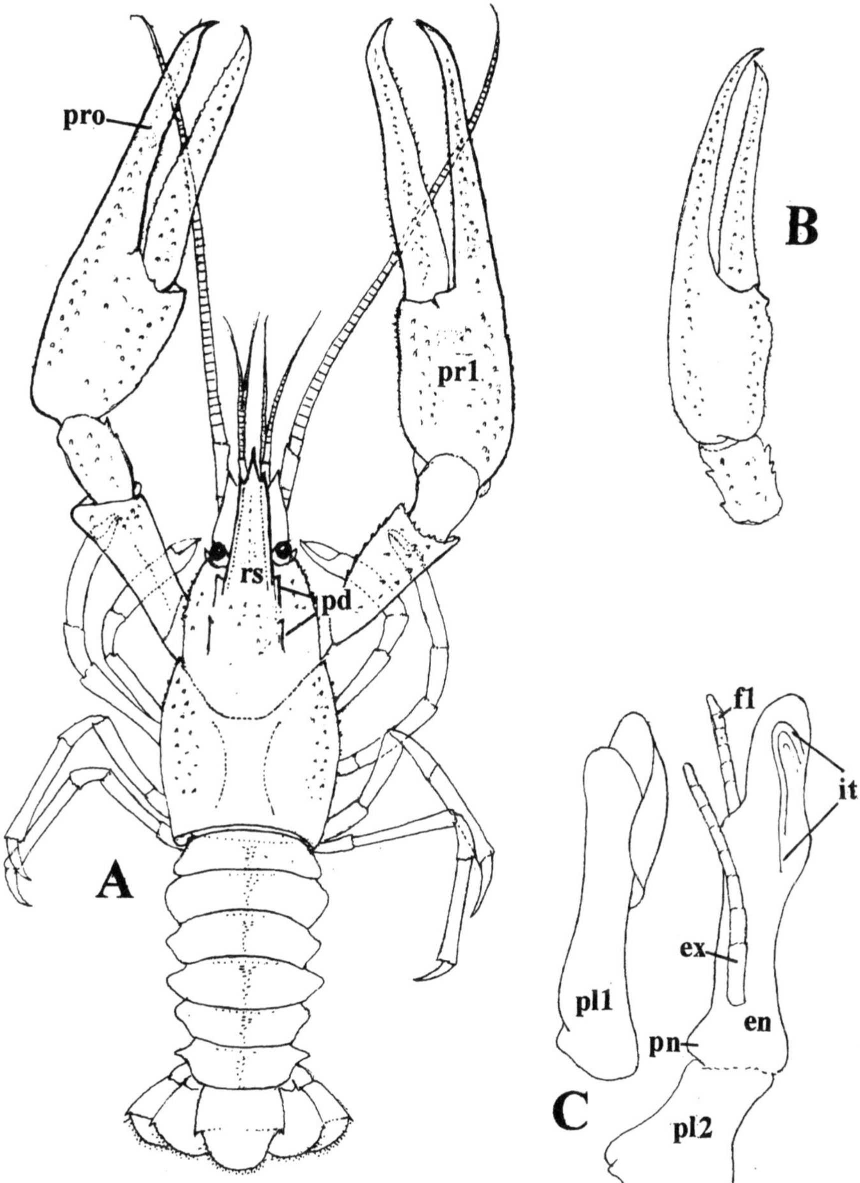

Fig. 4.6 *Astacus leptodactylus leptodactylus*, Long-clawed Crayfish. A. Dorsal aspect of whole crayfish. B. *A. leptodactylus salinus*, left cheliped. C. First and second pleopods of male from right side of body and from ventral aspect. For explanation of annotations see opposite page.

FURTHER READING

Anon. *A guide to identifying Freshwater Crayfish in Britain and Ireland.* National Rivers Authority (Information leaflet).

Alderman, D.J and Wickins, J.F. (1990) Crayfish culture. *Laboratory Leaflet* **62**. *Ministry of Agriculture Fisheries and Food.* Directorate of Fisheries Research.

Albrecht, H. (1982) Das System der europäischen Flußkrebse (Decapoda, Astacidae): Vorschlag und Begründung. *Mitteilungen aus den Hamburgischen Zoologischen Museum und Institut. Hamburg,* **79**, 187-210.

Bott, R. (1950) Die Flußkrebse Europas (Decapoda, Astacidae). *Abhandlungen hrsg. von der Senckenbergischen Naturforschenden Gesellschaft. Frankfurt a. M,* **483**, 1-36.

Curra, R.A. (1967) A key to genera, species and subspecies of Astacinae (Nephropsidea, Astacidae). *Internationale Revue der Gesamten Hydrobiologie. Leipzig,* **52**, 793-800.

Goddard, J.S. and Hogger, J.B. (1986) The current status and distribution of freshwater crayfish in Britain. *Field Studies,* **6**, 383-396.

Hart, C.W. Jr. and Clark, J. (1989) *An Interdisciplinary Bibliography of Freshwater Crayfishes (Astacoidea and Parastacoidea) from Aristotle through 1985.Updated through 1987.* Smithsonian Institution Press, Washington & London.

Holdich, D.M. and Lowery, R.S. (eds). (1988) *Freshwater Crayfish, Biology, Management and Exploitation.* Croom Helm, London & Sydney.

Laurent, P-J. and Forest, J. (1979) Données sur les écrevisses qu'on peut rencontrer en France. *La Pisciculture Française,* **56**, 25-40.

5

Marine Clawed Lobsters (Nephropoidea)

5.1 NATIVE SPECIES

All the Clawed [or True] lobsters mentioned in this book belong to the family Nephropidae. Those assigned to the genus *Homarus* are often termed homarid lobsters whereas those belonging to other genera, particularly *Nephrops* and *Metanephrops* are usually cited as nephropid lobsters.

Three species of clawed lobsters inhabit the northeastern Atlantic Ocean. The homarid European Clawed Lobster (*Homarus gammarus*), the nephropid Norway Clawed Lobster (*Nephrops norvegicus*) and the nephropid Scarlet Clawed Lobster (*Nephropsis atlantica*). The first two mentioned species are common and extensively fished commercially and the European Clawed Lobster is experimentally cultured. The Scarlet Clawed Lobster is occasionally captured by fishermen but is of no commercial value.

5.2 COMMERCIALLY IMPORTED SPECIES

The homarid American Clawed Lobster (*Homarus americanus*) supports extensive fisheries off the Atlantic coast of Canada and USA. It is regularly exported to Europe and the UK and is also experimentally cultured.

The relatively small quantities of imported nephropid lobsters at present all come from the Indo-Pacific region and belong to the genus *Metanephrops*. Features are given in table 11.1 (pp 196-197) for identifying these imported lobsters.

5.3 LIFE HISTORY

5.3.1 Homarid lobsters

Females of the European Clawed Lobster mature at about 25 cm in length. Mating occurs shortly after the female has moulted but before the shell has hardened. The male deposits sperm packets [spermatophores] into the female's sperm receptacle (**rp,** fig. 1.1, D) where they can remain viable for at least fifteen months. Spawning usually occurs during early autumn. The small fertilised eggs are incubated attached to oosetae (**os**, fig. 1.2, B) on the pleopods. Between 10,000 to 100,000 eggs may be spawned depending upon the size of the female. Throughout a period of incubation of nine to ten months the female preens the egg mass to remove debris and also maintains a rhythmic beating of the pleopods to oxygenate the eggs.

The hatched larvae [zoeae] emerge in late spring and early summer.Each is small and shrimp-shaped (fig. 5.1, A) and swim in the upper and mid-water off-shore region inhabited by the plankton community. Throughout the following six weeks the larva will moult three times or more increasing in size at each moult. At the third moult it is transformed into a small lobster shape form, the megalopa (fig. 5.1, B), that migrates to the sea bed occupying similar habitats to the adult, living in holes, crevices and other suitable hides. After the following moult it becomes a juvenile lobsterling.

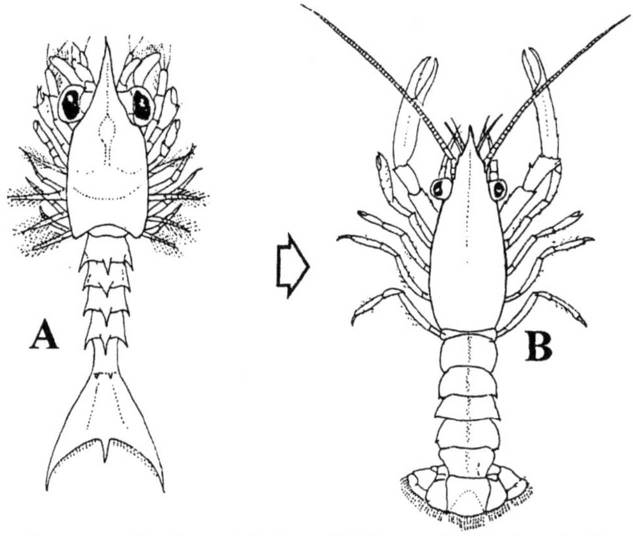

Fig. 5.1 Development of a homarid clawed lobster. A. Planktonic first stage larva [zoea], total length 6 mm. B. Benthic larval stage [megalopa], total length 20 mm.

5.3.2 Nephropid Lobsters

Females of Norway Clawed Lobsters mature at about 8.5 cm in length. At this size they can carry about 500 large eggs that increases to about 5000 for lobsters of 15 cm length. The majority of females spawn in early autumn and incubate their eggs for approximately eight months. Each hatches as a minute spiny larvae [zoea] (fig. 5.2, A) that moult through at least three planktonic stages before developing into a lobster-shaped [megalopa] form (fig. 5.2, B) that adopts a benthic life style on the sea bed. At the following moult it is transformed into a juvenile lobsterling.

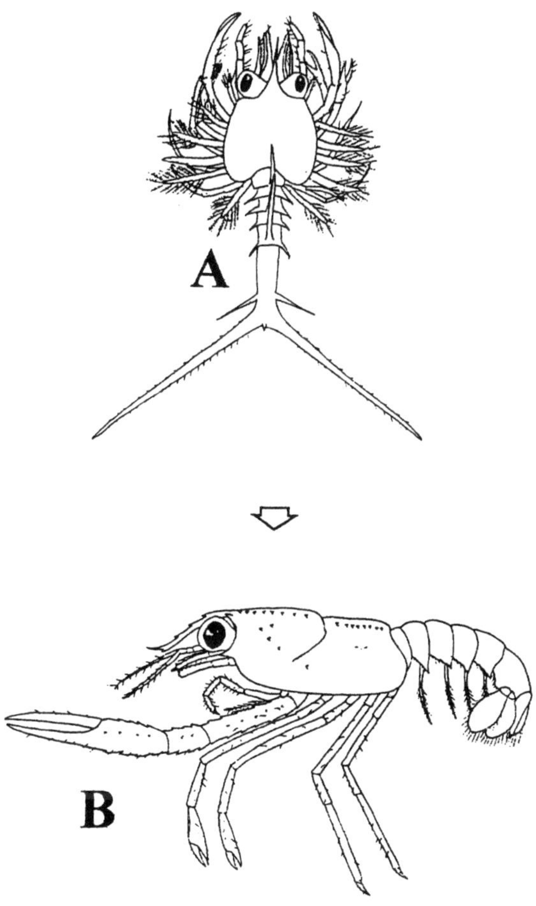

Fig. 5.2 Development of a nephropid lobster. A. Planktonic first stage larva [zoea], total length 7 mm. B. Benthic larval stage [megalopa], total length 1.5 cm .

5.4 FISHERIES FOR HOMARID LOBSTERS

5.4.1 Historical aspects

Although lobsters have been fished from the earliest times, it would seem that clawed lobsters did not become a 'fashionable' food item until the mid-eighteenth century. In the Middle Ages shellfish were considered as taboo by many communities and some countries forbade the fishing for lobsters as it was believed that their unrestricted presence on cod and herring grounds encouraged the appearance of these fishes.

The European Clawed Lobster, *Homarus gammarus* was at one time plentiful in shallow coastal waters. As late as the mid-eighteenth century they were collected with long wooden tongs of 3-4 m in length that were used in waters not exceeding about 3.5 m depth. Some lobstermen used a hooked staff in the form of a shepherds crook. This was thrust into crevices in which lobsters hid at low tide. Others used baited hoop nets. British populations of lobsters however, may never have been as abundant as those of the American Clawed Lobster, *Homarus americanus.* These were apparently used by american farmers in the 1850's '...by their thousands to manure the lands...'

By the mid-nineteenth century shallow water populations of European Clawed Lobsters had declined, no doubt due to overfishing, and it was necessary to fish in deeper waters using baited pots and traps specially constructed for catching lobsters. The overall design of these creels and pots has not changed appreciably throughout the last 200 years.

The British market for the European Clawed Lobster was at one time supplemented by imports from Norway whose fisheries were well organised for meeting export demands. As early as the seventeenth century the Dutch were buying lobsters from Norway and by the beginning of the eighteenth century the English were competing as serious buyers. From 1821-1824 more than one million lobsters were annually exported to England. This number had risen to an impressive one and a half million by 1828 after which time catches began to declined and prices were increased.

The Norwegians anticipated that demand for lobsters would eventually exceed supplies and formed the 'Stavanger Association' whose function attempted to limit and control the sale and export of lobsters from Norway. As early as 1737 it had been proposed that lobsters should not be fished during the breeding and moulting periods. But it was not until 1848 that a closed season was enforced in Norway when the English Lobster Company announced their intention to buy lobsters throughout the summer months. All regulations, particularly those enforcing a summer closed period, were generally strongly opposed by the majority of UK companies. In 1849 after more than 20 years of controversial debate, legislation was introduced by the Norwegians that forbade fishing of lobsters from 15th July until 30 September. In 1893 this closed period was further extended to 6th October. In 1922 these closed season dates were again ratified and a minimum landing size for lobsters of 21 cm total length was introduced for Norwegian waters and increased to 22 cm in 1964. Fishing regulations and minimum landing sizes now vary slightly from country to country. A closed season has never been enforced in UK waters where bye-laws used to regulate the minimum legal landed size that eventually became applicable to all UK waters. This was increased from 20.3 cm to 22.8 cm total length in 1951. The method of measuring was changed in 1976 to a carapace length of 8.0 cm, increased to 8.30 cm in 1981 and to the current length of 8.50 cm. in May 1984

Traditional fishing grounds in the UK for the European Clawed Lobster are the coasts of east Scotland, north eastern England, Wales and south east Ireland. The majority of European countries also support fisheries for this lobster and new populations are continually being sought. In all countries however, numbers of lobsters landed have declined since about 1975 with a corresponding increase in prices.

5.4.2 Fishing gear

The types of traps used to catch lobsters are constructed around two basic designs that vary slightly from country to country. In the United Kingdom the *Cornish or Inkwell Pot* (fig. 5.3, A) is used chiefly by lobstermen fishing the south west coasts of England and Wales. The frame (**fr**) is sometimes constructed of wickerwork but galvanised wire and netting is now commonly used. An average Cornish pot measures 1-1.25 m in diameter but larger ones are employed for storing lobsters. The pot has a single top opening (**op**) or 'eye' of 20-25 cm diameter that is often protected by two or three skewers projecting inwards towards the centre. These prevent lobsters escaping and are often used for attaching the bait. Stone weights (**wt**) are secured to the base of the pot.

The frame of the *creel* (**fr**, fig. 5.3, B) is cylindrical or sub-oval and of variable design but has one or more openings (**op**) or 'eyes' on the sides. Variously named local designs, (eg. Norfolk Creel-north east coast of

England and the Parlour Creel-Scottish coasts) made from oak or spruce covered with wood lathes, has almost disappeared. Many lobstermen now use commercially made metal frame creels covered with nylon netting, these can be conveniently collapsed for storage. Nylon netting reduces the creel's buoyancy and less ballast is required for the anchor. A barrel-shaped creel is still used by French lobstermen. Creels vary in size but average 1 x 1.5 m with entrances of 10-12 cm diameter. These entrances, also termed 'heads', vary in position according to creel shape and locality. Various experimental designs have been tried to improve catch efficiency. One commercial creel has a row of bristles arising from the floor over which the lobster passes as it enters the trap and which prevents escape. Another escape prevention device is an inverted U-shape piece of galvanised wire hinged across the inner opening of the 'eye' and while offering little resistance to entry of large lobsters may deter small lobsters from entering the trap. Experimental evidence suggests that the Cornish Pot is just as efficient as the creel for catching lobsters and that creels with a 17 mm mesh fitted to the inner lower eye catch more lobsters than creels using a large mesh.

5.4.3 Bait

Bait used for luring lobsters into pots and creels varies from region to region. Fish heads are favoured, particularly those of low commercial value; limpets and mussels are also occasionally used. In the past sea birds such as Shags, Comorants and Guillemots were caught, skinned, salted and used for bait.

The bait is impaled on skewers in Cornish Pots and placed in a bait box or bag in creels. The latter method protects the bait from being consumed by crabs and starfishes and occasionally by Sea Lice (*Conilera cylindracea*), a small carnivorous isopod crustacean that in large numbers can devour bait within hours. Experiments have not confirmed the widely held belief that stale bait is more effective than fresh bait.

5.4.4 Fishing methods

The peak fishing season in British waters is from May-October. In most regions trapping is the only practical method for catching lobsters in any quantity although in parts of the Mediterranean they are caught using trammel nets (a form of tangle net).

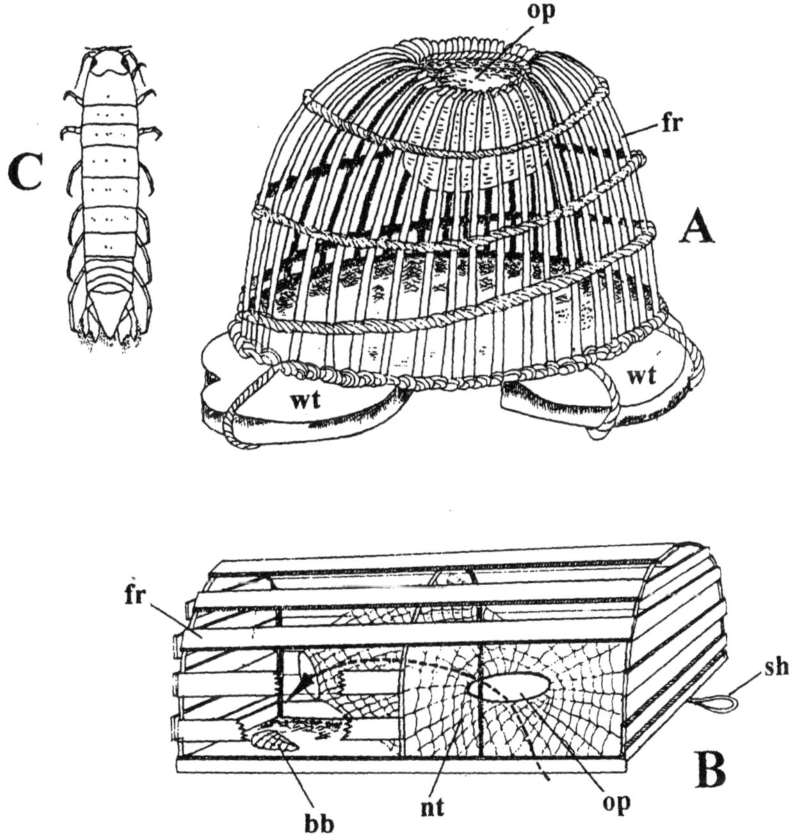

Fig. 5.3 These illustrations of a pot and creel used for catching homarid lobsters have been simplified for structural clarity. A. Inkwell or Cornish Pot. B. Parlour Creel, nylon nettings is now usually used in place of wood slats. C. The small bait-eating Sea Louse, length 17 mm: **bb** bait bag; **fr** framework; **nt** netting; **op** opening or 'eye'; **sh** shackle for attaching mooring rope; **wt** weights. Broken arrow shows path taken by lobster to reach the bait.

In UK waters small boats 4-6 m in length and powered by outboard motors are typically used for coastal fishing. Boats of this size can be managed single handed and do not usually carry more than 20-30 creels or pots that are 'worked' in waters usually not exceeding 19 m depth. However, fishing from such small vessels is much dependant on suitable weather conditions. Larger boats of 7-14 m in length with inboard engines can operate over a much wider area and for longer periods. They carry many more creels (often in excess of 100) that are laid in waters of 20-30 m depth.

Pots and creels are usually 'shot' singly from small boats and in fleets from large vessels. The pots are often baited during the outward journey to the fishing ground. In the method of 'single shooting' a float, attached to the creel by a rope, is cast overboard followed by the creel. Expert manipulation of the rope ensures that the creel is guided onto the correct place on the sea bed and arrives seated on its base. This operation is repeated for each creel.

Large vessels manned by two or three lobstermen may carry 200 creels or more and use the 'fleet setting' method shown in fig. 5.4. Creels are shot in batches of 8-12 and each is attached to the leader (**ld**) or ground rope by a mooring rope (**mr**) of 5-6 m in length. Mooring ropes are spaced along the leader at distances of 9-18 m. To each end of the leader is attached a heavy weight (**wt**) and buoy rope (**br**); the length of the latter is regulated by the water depth. Corks (**ck**) are attached to the buoy ropes at approximately 2 m intervals to prevent these ropes fouling the sea bed at low tide. Each buoy rope is attached to a buoy (**bu**) made of painted wood or plastic and these mark the locations of the set creels.

Creels are fleet set with or across the tide with the boats moving forward; setting against the tide would cause the creels to pile on top of each other. Creels and pots are usually left overnight and lifted the following morning. Working with the tide reduces the considerable strain of hauling. Early morning lifting will catch that day's market or transport if lobsters need to be sent long distances. Late hauling, often due to bad weather, means that lobsters have to be stored alive, usually in creels, until they can be marketed. It is common practice to rebait and reset creels at the morning's lifting. In regions where lobsters are reasonably abundant, for example off parts of the coasts of Scotland, creels may be lifted twice or three times each day depending on tidal and weather conditions. Fleet hauling is strenuous work and is now usually assisted by a powered winch. The weight or anchor is hauled along with the first leader rope and buoy. These are stowed in a roller box usually located near the stern. A turn or two of the free leader is then wound around the winch capstan and as each creel is hauled its mooring rope is detached and coiled into the leader box. The creels are arranged along the deck in order of hauling. This operation is continued until the second leader, weight and buoy is retrieved.

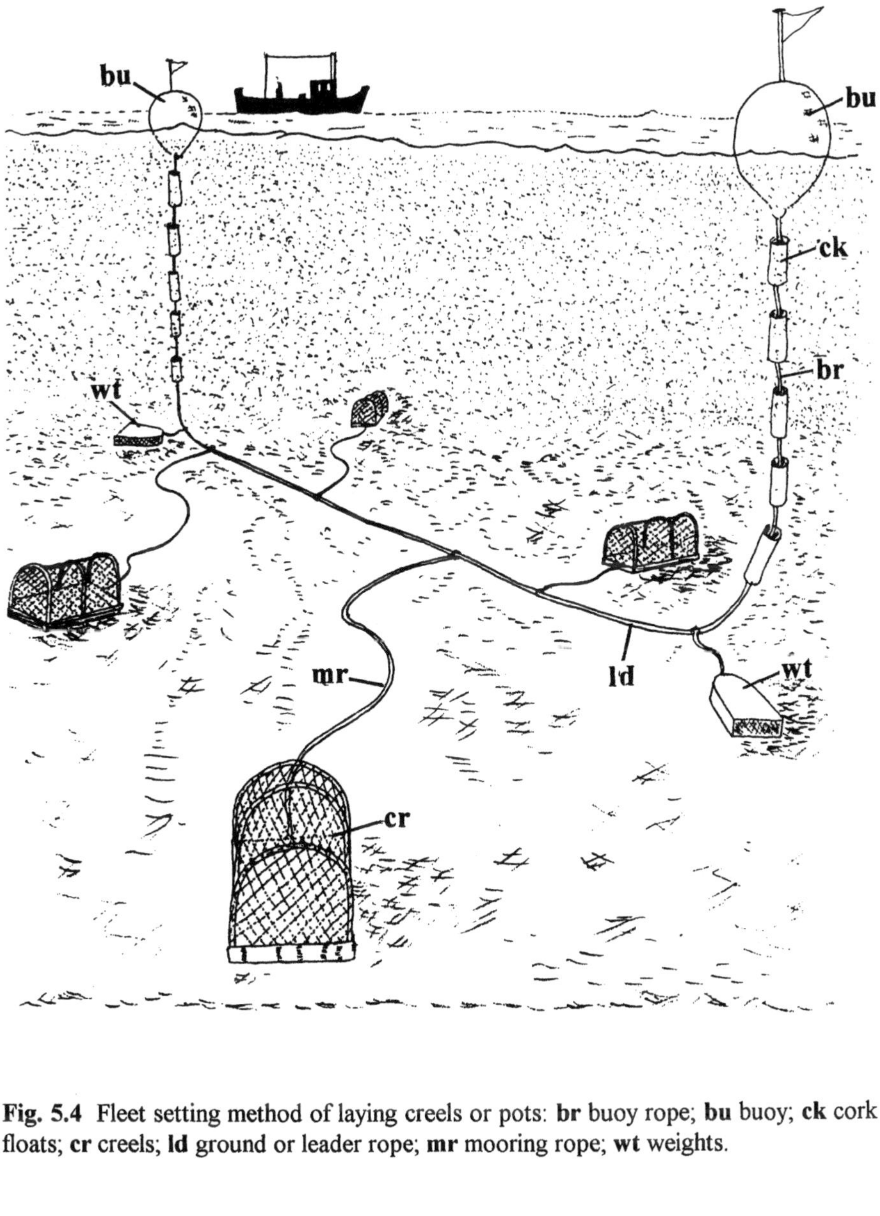

Fig. 5.4 Fleet setting method of laying creels or pots: **br** buoy rope; **bu** buoy; **ck** cork floats; **cr** creels; **ld** ground or leader rope; **mr** mooring rope; **wt** weights.

5.5 FISHERIES FOR NEPHROPID LOBSTERS

5.5.1 Historical

The nephropid lobster sold in quantity on UK and European markets is the Norway Clawed Lobster (*Nephrops norvegicus*) often cited as the Dublin Bay Prawn. This lobster was not fished in any significant quantity until about the 1950's although from the early part of this century fishermen were aware that large numbers could be easily caught as trawling became more efficient. During the years immediately after World War II, Norway Clawed Lobsters caught off the coast of Scotland by vessels trawling for white fish, were only occasionally landed. Where there was a market demand from some ports these lobsters were then landed and sold as ancillary to fish catches. However, during the last half of this century the landings of Norway Clawed Lobsters by many European countries increased considerably creating highly profitable fisheries. France still lands the largest quantity, followed by Scotland, Spain, Ireland, Iceland and Italy.

UK fisheries are traditionally located in Scottish and south western Irish waters. French landings come chiefly from the Bay of Biscay where a special beam trawl fisheries was developed, but French vessels also fish parts of the southern North Sea and Irish Sea. Scandinavian countries fish mostly in the Skagerrak and Kattegat regions. This latter area and the southern North Sea is fished by German vessels particularly during the summer but less so during the winter when the Lobster forms only an incidental part of white fish catches. Icelandic fisheries are located chiefly off the south and south west coasts of Iceland and in waters of more than 100 m depth. In the Mediterranean the major fishing regions are the Adriatic and Tyrrhenian Seas.

5.5.2 Fishing methods

The largest populations of Norway Clawed Lobsters occur in relatively deep water and live for most of the time in burrows that they make in soft bottom substrates. It was discovered that they emerge to forage for food at times approximating to sunrise and sunset and this knowledge improved the catching of quantities of these lobsters.

The chief method for fishing is by trawling. About 1956 the small mesh seine net used to trawl these lobsters was modified into what became known as the standard 'Nephrops Trawl' that is used in regions where trawling is permitted.

Fishing regulations pertaining to minimum landed sizes, net mesh diameters, methods of fishing and fishing vessel sizes differ from, and even within, one country to another. In UK waters originally a net with a 50 mm diameter stretched mesh was used, but in 1962 in order to conserve stocks, a Sea Fishing Order was enforced making it illegal to use a net that had a mesh smaller than 70 mm. (See p. 58 for current regulations).

In regions where the sea bed is unsuitable for trawling or where trawling is prohibited by law, baited creels are used. These are of similar construction to those described for catching homarid lobsters.

5.6 MARKETING CLAWED LOBSTERS

5.6.1 Homarid lobsters

Catches of the European Clawed Lobster are marketed whole. The condition of lobsters when landed affects the wholesale price. Care is given therefore, to proper storage and packing. Undamaged specimens of approximately 0.5 kg to 0.75 kg appear to command the best prices in UK markets.

Lobsters are usually transported alive. Each of the paired large claws [chelae] are tied [banded, fig. 5.5] to prevent lobsters damaging each other; if time allows this is sometimes done on the inward journey from the fishing grounds. Packing is aimed at keeping respiration to a minimum so that lobsters arrive alive after being transported to inland markets or to other purchasers such as hotels and restaurants. They are usually packed (after grading into small, medium, large and mutilated specimens) into long wooden boxes, close together, back uppermost and with abdomen bent beneath the thorax and claws projecting forward. Wet sacking is sometimes used to cover boxed lobsters before transport by road or rail. In warm weather ice cubes may be used. There are placed upon a layer of waxed paper to prevent melting freshwater from reaching the lobsters. Final packing weight of each box is usually related to freight charges at the time of transport.

In the UK and most European countries lobsters are generally cooked as soon as possible after purchase by which time they are usually moribund. Deep freezing uncooked lobsters is now not usually recommended although flavour and quality is considered by some to be better than frozen cooked meat; the latter is often sold for salad preparation.

The large commercial concerns boil lobsters in rectangular tanks or vats. These are often only partially filled with water and injected with steam. When the correct temperature is reached the then moribund lobsters are placed into baskets and lowered into the tanks. Salt is added to the water and the lobsters boiled until they assume the correct red colour indicating they are sufficiently cooked. Some may be partially dismembered and machine packed for sale to food retailers whereas the meat from others is used to prepare lobster paste.

Lobsters not immediately marketed are stored alive either in large creels or pots moored in sheltered coves, harbours, or in land-based tanks and tidal ponds. In the UK and Europe the holding of lobster in ponds has not been developed on the large scale seen in North America. Nevertheless, some European ponds can successfully hold 20,000 lobsters incurring low mortality rates of about 10%. Lobsters held in this manner must have their claws banded to prevent cannibalism. In these holding ponds, bottom layers of water can quickly become depleted of oxygen and must be pumped to the surface and aerated by compressed air; alternatively, water from the sea is allowed to flow freely into the pond. Lobsters stored in ponds for more than six weeks are usually fed when the water reaches summer temperatures of about 15° C; at temperatures below 5° C lobster do not feed. When required for sale the lobsters are removed from ponds by drag nets, baited hoop nets or by draining the pond to a depth at which they can be caught with a dip net.

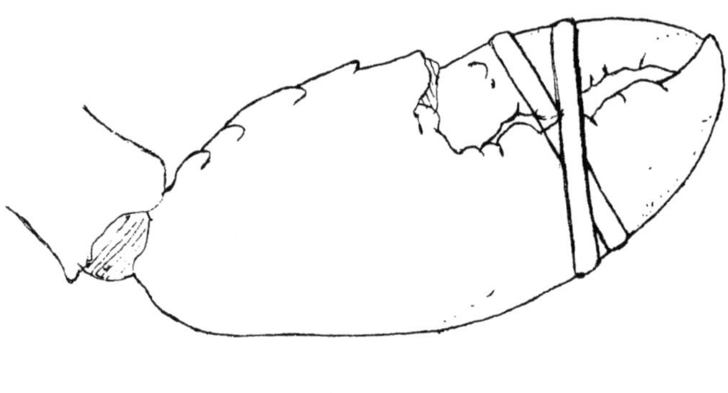

Fig. 5.5 The immobilised [banded] claw of a lobster. Stout rubber bands are sometimes used.

5.6.2 Nephropid Lobsters

Although a large part of Norway Clawed Lobster catches are marketed whole, the practice of selling only the 'tails' has now become more widespread. Vessels fishing for these lobsters are often away from ports 7-8 days at a time. To reduce deterioration of the meat it is usual practice to wash the catch on deck and to quickly transfer it as near to freezing temperatures as possible. Often on long voyages only the 'tails' are retained for icing as the 'heads' are a source of spoilage. Stowing lobsters and fish in the same hold is now not recommended.

On return to port catches are immediately unloaded and sold. On some quaysides the lobsters may be graded as 'big' (less than 40 per kg), 'medium' (40-50 per kg) and 'small' (more than 50 per kg). 'Tails' are similarly graded. The catches are then placed into baskets, hosed and taken to the auction hall. Each basket can contain 30 kg of lobsters. After auction most are cooked and distributed through wholesale or retail outlets.

Cooking involves a further washing to eliminate mud, sand and slime that would soil the cooking water. Cooking methods vary from plant to plant. In addition to salt, spices are sometimes added to the water to improve taste of the meat; red pepper, thyme, bay leaves, celery and onions are also used. When the water is boiling lobsters are added by hand or the baskets in which they have been washed is lowered into the cooking vat. Cooking continues until the lobsters have assumed a pink/red colour after which any scum is removed before they are packed for distribution.

5.7 AQUACULTURE OF HOMARID LOBSTERS

For many years it has been accepted by those managing or advising on lobster fisheries that stocks of both the homarid European (and American) clawed lobsters will continue to seriously decline if the present level of fishing intensity is maintained. This situation had been realised even during the early part of the nineteenth century and stimulated the Norwegians to develop methods for culturing the European Clawed Lobster. Since that time much effort has been expended on developing methods for reliable and profitable commercial rearing of both this species and the American Clawed Lobster.

Lobsters breed successfully in captivity, produce large numbers of eggs, have a short larval life and can be reared to a post-larval stage without too much difficulty. However, development of eggs from fertilization to hatching unfortunately takes many months and post-larval growth is slow; it requires two or more years for juveniles to reach a marketable size. More seriously is the trait of lobsters to extremes of aggressive behaviour towards each other and a preponderance towards cannibalism. Both of these undesirable features become very evident when lobsters are crowded in culture conditions. Capital investment in lobster aquaculture can be high with no financial returns for two or more years. Nevertheless, experimental commercial culture continues in many European countries. Some of these efforts have been directed at producing juveniles for local restocking of fished regions, others at developing commercially feasible growing-on methods, such as using heated waste water from industrial plants or electricity generating stations. At present however, there seems little long-termed financial reward in commercial rearing of lobsters. Current methods are still labour intensive and the true costs are not known of growing on sufficiently large numbers of juveniles to a size that will regularly supply market demands.

Rearing methods vary from one hatchery to another. Essentially, female lobsters are allowed to hatch their eggs in specially designed containers through which water flows carrying hatch larvae into collecting baskets. The larvae are then transferred to rearing vessels containing water in gentle turbulence that keep them moving to reduce cannibalism and to ensure even distribution of food. Although ground fish, shellfish or artificial diets are used to feed the larval stages, the most nutritious food that gives the highest survival rates is adult or late larval stages of the Brine Shrimp (*Artemia*). When the larvae moult to post-larvae they are then usually reared in individual growing-on containers. One juvenile in each container has proved the most satisfactory method for obtaining uniform growth rates and low mortalities. As the lobsters grow chopped fish and shellfish is gradually substituted for the brine shrimp diet.

5.8 IDENTIFICATION

The three species of nephropid clawed lobsters (subfamily Nephropinae) occurring in North Eastern Atlantic waters are easily identified one from the other. The European Clawed Lobster, *Homarus gammarus* has conspicuously broadened chelae (**ce**, fig. 5.6) these, and the body surface, are smooth. By comparison the Norway Clawed Lobster, *Nephrops* *norvegicus* has narrow spinose and ridged chelae (**ce**, fig. 5.7) and the abdominal segments (**ab**) have transverse grooves.

The rarely caught Scarlet Clawed Lobster, *Nephropsis atlantica* is recognised by the very small unpigmented eyes and the numerous hairs covering the body surfaces.

Homarus gammarus (Linnaeus, 1758)
Fig. 5.6, Pl.9.
COMMON SYNONYMS: *Homarus vulgaris.*
DERIVATION: homar (old French)-a lobster; kammaros (Greek)-a kind of lobster.
VERNACULAR NAMES: European Clawed Lobster, Common Lobster (UK); Hummer (Dn, Sc, Gn); Kreeft (Nd); Homard (Fr); Bogavante (Sp); Lavagante (Pt); Astice (It); Astakos (Gr); Istakoz (Tr).
DISTINCTIVE FEATURES: carapace (**cr**) smooth, rostral (**rs**) lateral margins with 4-5 teeth, lower margin without teeth, medial groove (**mv**) present throughout carapace length; abdominal segments (**ab**) smooth; first pair of pereiopods (**pr1**) smooth, chelae (**ce**) dissimilar, moderately broadened, (one more than the other), one with irregular large teeth, other narrow and with small often sharper teeth.
COLOUR: bluish to almost black, with lighter reticulations, underside white to yellowish.
TOTAL LENGTH: commonly 35-40 cm; rarely exceeding 50 cm. (Largest known 62 cm and weighing 8.4 kg).
DISTRIBUTION: North Eastern Atlantic-Norway to Morocco. Whole of Mediterranean including Black Sea.

The European Clawed Lobster inhabits inshore low water downwards to 60 m depth and occurs among rocky/stony substrates.
GENERAL INFORMATION: females bear eggs from early autumn and developing young until April or May of the following year.

The lobster is fished in coastal waters by pots, creels, and (in the Mediterranean) by trammel nets.

Minimum legal landing size in UK waters is 22.5 cm.

The lobster is experimentally cultured.

To import or move live or dead Clawed Lobsters within coastal regions of the British Isles a licence is required from the Ministry of Agriculture Fisheries and Food.

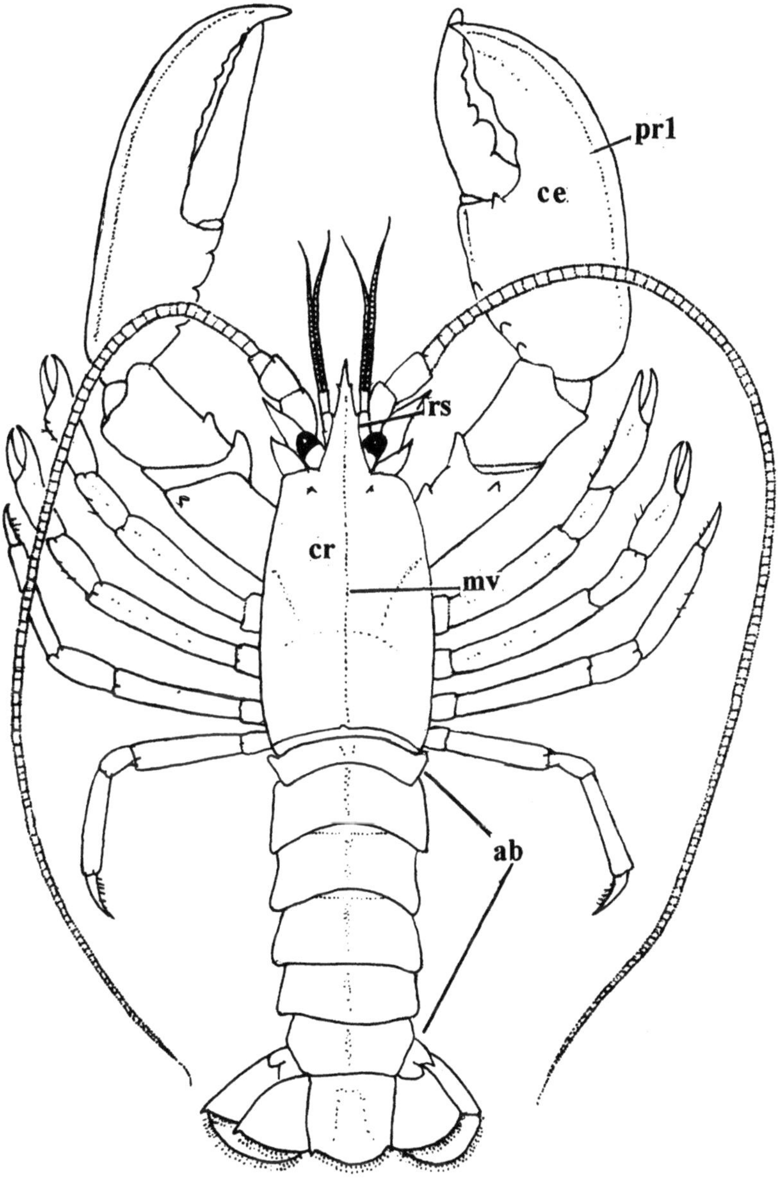

Fig. 5.6 *Homarus gammarus*, European Clawed Lobster. For explanation of annotations see opposite page.

Nephrops norvegicus (Linnaeus, 1758)
Fig. 5.7, Pl. 11.
COMMON SYNONYMS: none.
DERIVATION: nephros (Gr)-a kidney, pertaining to shape of eyes; norvegicus (Lt)-pertaining to Norway.
VERNACULAR NAMES: Norway Clawed Lobster, Dublin Bay Prawn, Scampi (UK); Dybvandshummer (Dn); Sjøkreps (Nw); Havskräfta (Sw); Keisarihummeri (Fi); Noorse Kreeft (Nd); Kaiserhummer (Gn); Langoustine (Fr); Cigala (Sp); Lagostim (Pt); Scampo (It); Karavida (Gr); Nefrops (Tr).
DISTINCTIVE FEATURES: carapace (**cr**) with longitudinal rows of sharp spines anteriorly, rostrum (**rs**) thin, lateral margins with 3-4 spines; dorsal surfaces of abdominal (**ab**) segments with broadened hairy grooves; chelae (**ce**) of first pair of pereiopods (**pr1**) more or less of equal size, narrow and with longitudinal spinous ridges.
COLOUR: pinkish red, underside pale.
TOTAL LENGTH: about 20 cm, exceptionally 25 cm.
DISTRIBUTION: North Eastern Atlantic-Iceland to Morocco. Whole of Mediterranean, abundant in Adriatic.

The Norway Lobster occurs chiefly offshore, in depths between 100-300 m and inhabits burrows in sandy/mud substrates.

GENERAL INFORMATION: females bear eggs from early autumn, these hatch in May, or later, of the following year.

The lobster is fished with bottom trawls in most regions but with creels in some areas. It is marketed fresh, but in the UK 'tails' often processed and sold frozen.

The UK Food Labelling Regulations of 1984 specify that the only crustacean permitted to be labelled and sold as 'Scampi' is *Nephrops norvegicus*.

Minimum landed sizes of Norway Clawed Lobster are governed by European Union Council Regulations. Landed sizes vary from 7 cm total length in the Irish Sea and waters of western Scotland, to 13 cm for lobsters fished in the Skagerrak and Kattergat regions of Scandinavia. The fishery is also regulated by permitted net mesh diameters. A minimum mesh of 70 mm is allowed for fishing *Nephrops* in UK waters but further southwards nets with mesh sizes of 55-65 mm are allowed, depending upon the type of trawl.

The name Dublin Bay Prawn is derived from the once time practice by fishermen of selling incidental catches of these lobsters landed at Dublin Bay (Ireland) to local street vendors, including no doubt the Molly Malone epitomised in the well know song.

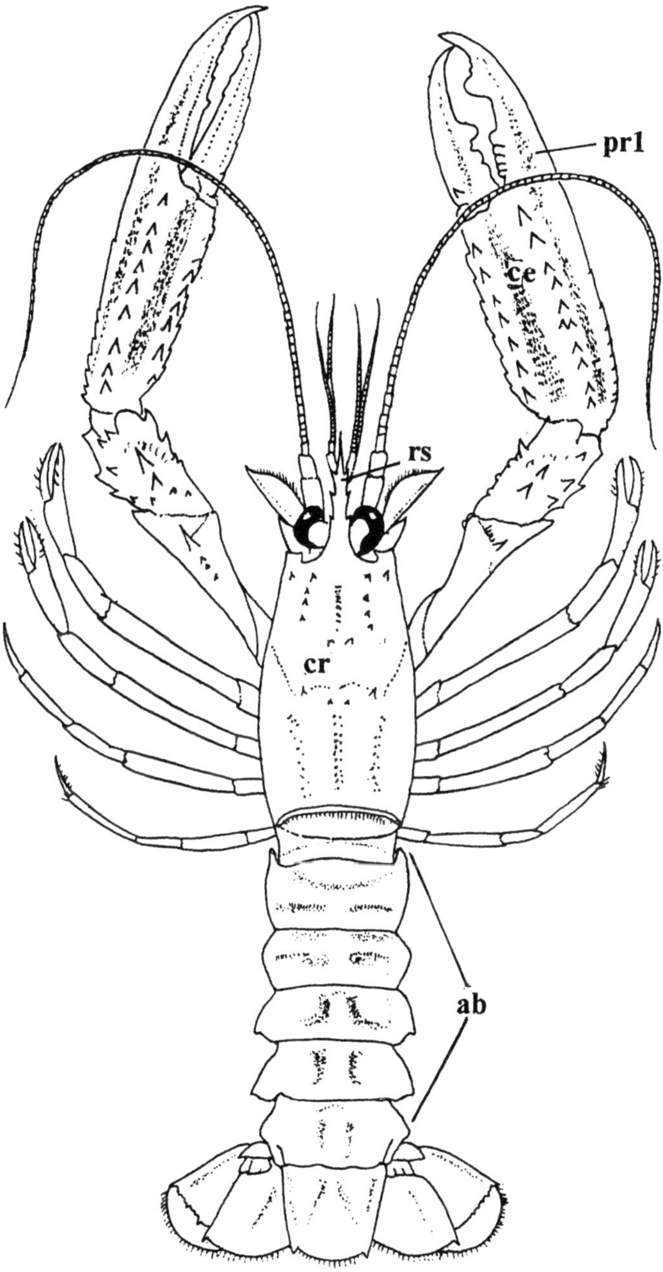

Fig. 5.7 *Nephrops norvegicus*, Norway Clawed Lobster. For explanation of annotations see opposite page.

FURTHER READING

Allen, J.A. (1967) *The Fauna of the Clyde Sea Area. Crustacea: Euphausiacea and Decapoda.* Scottish Marine Biological Association, Millport (pages 56 & 88 for identification of clawed lobsters).

Cobb, J.S. & Phillips, B.F. (edits). (1980) *The Biology and Management of Lobsters. Vol. I. Physiology and Behaviour. Vol. II. Ecology and Management.* Academic Press, New York, London.

Davis, F.M. (1958) An account of the fishing gear of England and Wales. *Fishery Investigations. Board (Ministry) of Agriculture Fisheries and Food.* London, (II) **XXI**, 1-165.

Figueiredo, M.J. and Thomas, H.J. (1967) *Nephrops norvegicus* (Linnaeus). A review, in Barnes, H (ed). *Marine Biology Oceanography Annual Review*, **5**, 371-407.

Holthuis, L.B. (1974) The lobsters of the superfamily Nephropidae of the Atlantic Ocean (Crustacea: Decapoda). *Bulletin of Marine Science, Gulf of Caribbean*, **24**, 723-884.

Holthuis, L.B. (1991) Volume 13, Marine Lobsters of the World. FAO Species Catalogue. *FAO Fishery Synopsis*, **125**, FAO Rome, (pages 19-86 for identification of clawed lobsters).

Zariquiey Alvarez, R. (1968) Crustáceòs Decápodos Ibéricos. *Investigación Pesquera,* **32**, 1-510, (pages 198-202 for clawed lobsters).

6

Marine Crawfishes and Shovel-nosed Lobsters (Palinuroidea)

The term crawfish is used here for all lobster-shaped decapods belonging to the family Palinuridae. These are often cited as 'spiny lobsters' and sometimes as 'marine crayfishes'. In this book the vernacular name 'spiny lobster' is applied only to species belonging to the genera *Palinurus* and *Panulirus*. Similarly, the name 'shovel-nosed lobster' includes all species of the family Scyllaridae. The vernacular name 'slipper lobster' is restricted to species of the genus *Scyllarides* and 'locust lobster' to species of *Scyllarus*. The use of common names for these lobsters is further discussed in sections 11.3, 11.4.

Crawfishes (see fig. 3.1, C) have small or large spines covering the carapace. The rostrum is usually absent or, in a few species, reduced to a single spine (**rs**, fig. 6.2, A); a pair of frontal horns (**fh**) are usually conspicuously developed over the eyes. The flagellum (**fl**, fig. 3.1, C) of the second antenna is long and often whip-like. The abdominal pleura (**pu**) or 'side plates' terminate in acute backward or downward directed curved or straight processes.

By comparison, the carapace of shovel-nosed lobsters (fig. 3.1, D) has blunt to sub-acute spines and the rostrum or rostral horns are absent. The second antenna (**an2**) is modified into a flattened paddle-like plate and the abdominal side-plates terminate in irregular or granular margins that do not extend as acute processes.

6.1 NATIVE SPECIES

Three species of crawfishes inhabit the North Eastern Atlantic/Mediterranean regions. The Common Spiny Crawfish (*Palinurus elephas)* is frequently offered for sale in many fish markets and restaurants, but the Pink Spiny Crawfish (*Palinurus mauritanicus)* is far less often seen. The Royal Spiny Crawfish (*Panulirus regius)* has the limits of its northernmost range in the western Mediterranean and is more frequently encountered in the markets of this region than in those of more northern countries.

Five species of shovel-nosed lobsters occur in the North Eastern Atlantic/ Mediterranean regions. The Mediterranean Slipper Lobster (*Scyllarides latus*), and four species of locust lobsters, namely the Small European Locust Lobster (*Scyllarus arctus*-the only species species distributed to the southern British Isles), the Pigmy Locust Lobster (*S. pygmaeus*) and two species without vernacular names-*S. caparti* and *S. posteli*. In the Mediterranean *S. latus* and *S. arctus* are fished and whole lobsters command high prices perhaps because they are relatively scarce and the meat is of high quality. *S. pygmaeus* is sometimes misidentified by fishermen as juveniles of *S. arctus*.

6.2 COMMERCIALLY IMPORTED CRAWFISHES

A considerable number of crawfish species are fished throughout coastal waters of tropical and sub-tropical regions of the World. Most have high market values and there is an extensive trade in their frozen 'tails'. Many are imported without specific identifications but several, and at least one shovel-nosed lobster species, have been noted by the author offered for sale in the UK and many will continue to appear in European markets from time to time. These imported species are discussed in chapter 11.

6.3 FISHING METHODS FOR CRAWFISHES

Crawfishes are fished using tangle nets, creels, sub-aqua hand collecting and spear-gun fishing. In some regions they provide a financially rewarding bye-catch when trawling for fish or prawns. Usually once method predominates locally; commercial vessels often freeze 'tails' on board.

6.4 LIFE HISTORY

The life history summarised below for the European Spiny Crawfish is typical for all crawfishes but time of egg carrying and larval development varies according to the species.

The eggs are spawned in September and October and are incubated attached to the pleopod oosetae of the female. They hatch in May or June as bizarre transparent leaf-like larvae (fig. 6.1, A) termed phyllosomas. These moult through many stages of a similar shape, increasing in size at each moult acquiring additional appendages and body segments at each stage. During these planktonic stages, that may last many months, the larvae can drift considerable distances. The mouthparts remain relatively undeveloped and little is known about their food requirements. They are probably carnivores; the larvae of tropical species have been seen to feed upon jellyfishes. The last planktonic stage moults into a form (fig. 6.1, B), often termed a puerulus, that is adapted for living on the sea bed where it further moults into a juvenile acquiring most of the features of the adult. Shovel-nosed lobsters have a similar life history; the shape of the benthic stage (fig. 6.1, C) is very similar to the adult.

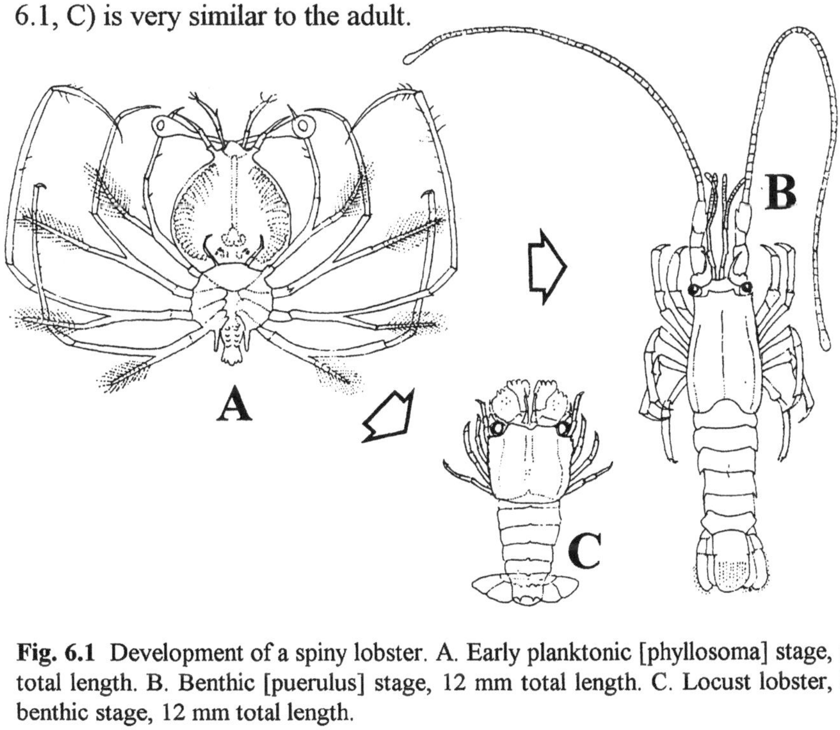

Fig. 6.1 Development of a spiny lobster. A. Early planktonic [phyllosoma] stage, 5 mm total length. B. Benthic [puerulus] stage, 12 mm total length. C. Locust lobster, larval benthic stage, 12 mm total length.

6.5 AQUACULTURE OF CRAWFISHES

The ever increasing market demand for crawfishes and the decline in numbers of some fished populations has initiated studies directed towards their commercial culture, particularly of spiny crawfishes.

Although a few species of spiny crawfishes and shovel-nosed lobsters have been successfully reared to juvenile stage, long larval life and insufficient knowledge of food and environmental requirement have presented formidable difficulties for rearing them by the methods used for clawed lobsters (5.7).

The leaf-like phyllosoma stage may last for as long as six months and there can be at least eleven stages of this form before the change occurs to the benthic puerulus stage. Growth is slow and laboratory experimental rearing suggest that at local sea temperatures juvenile crawfishes may take seven years to reach a marketable size but that this time can be reduced to two years if they are reared in high temperatures of about 22°C. However, crawfishes reared above ambient temperatures are more active and require to be fed more frequently. This, and other factors, inevitably incur higher maintenance cost. A further problem is the difficulty of obtaining sufficient numbers of pueruli or juveniles for 'rearing on'. One approach has been to devise special traps to collect post-larval juveniles and to release these in suitably managed holding ponds where they can be reared to acceptable sizes for sale.

Even when reliable methods are developed the high costs involved in commercial rearing would make the price of cultivated crawfishes uncompetitive until such times when natural stocks become virtually fished out or unless special provision was made for marketing these reared crawfishes at an agreed size below that of the legal local ones in force.

6.6 IDENTIFICATION

The European and the Pink Spiny Crawfishes (*Palinurus*) are easily recognised by the very short flagella (**f1**, fig. 6.2, A) of the first pair of antennae (**an1**) and the minute rostral spine (**rs**) present between the frontal horns (**fh**). The combined features distinguish these two species from the Royal Spiny Crawfish (or any other species of *Panulirus*) in which the antennular flagella are very long and the rostral spine absent.

The Mediterranean Slipper Lobster (fig. 6.4) has the margins of the antennae (**an2**) undivided and the fifth pair of pereiopods do not terminate in small claws. In both species of locust lobsters (Small European, and Pigmy) the antennal margins (**an**, fig. 6.5) are distinctly lobed and the fifth pair of pereiopods (**pr5**) terminates in a claw (**cl**) for females.

One of the two tropical Atlantic locust lobsters, *Scyllarus caparti* is known from the Mediterranean. The other species, *S. posteli* occurs northward as far as the Gulf of Cadiz (Spain). *S. caparti* has a high longitudinal dorso-medial crest on the abdominal segments and the posterior margins of these segments have 5-6 denticles. The posterior region of the thoracic sternum of *S. posteli* has a prominent anteriorly directed spine.

Palinurus elephas (Fabricius, 1787)
Fig. 6.2, Pl. 13.
COMMON SYNONYMS: *Palinurus vulgaris.*
DERIVATION: Palinuris (Gr)-steersman and Trojan sailor to Aenaeus; proverbially used for advisor or pilot; elephas (Lt)-an elephant, denoting elephantine size.
VERNACULAR NAMES: European Spiny Crawfish, Spiny Lobster, Rock Lobster (UK); Langust, Languster (Sc); Langusti (Fi); Langoest (Nd); Languste (Gn); Langouste Rouge (Fr); Langosta Común (Sp); Lagosta (Pt); Aragosta Mediterranea (It); Astakós (Gr); Bocek (Tr); Jarradh el Bahr (Tn).
DISTINCTIVE FEATURES: carapace (**cr,** fig.A) with numerous large and small spines, a small rostral spine (**rs**) between the frontal horns (**fh**); dorsal surface of abdominal (**ab**) segments never with flat, scale-like sculpturing, pleura (**pu,** & fig. B) with three to five small spines on hind margin; flagella (**fl**) of first antenna (**an1**) very short; first (**pr1**) to third (**pr3**) pereiopods without chelae; distal parts of telson (**tl**) and uropods (**ur**) fleshy; male-propodus (**pro,** fig. C) of first pereiopod with disto-external spine (arrowed).
COLOUR: brownish red to purple, spines darker; a pair of white to yellowish bands and dark spots on each abdominal segment; pereiopods with longitudinal brownish yellow patches or lines.
TOTAL LENGTH: 40 cm, rarely reaching 50 cm.
DISTRIBUTION: North Eastern Atlantic-from Hebridean region southward to Azores. Western Mediterranean to coast of Turkey.

The European Spiny Crawfish occurs on rocky substrates in depths of 20-70 m.

GENERAL INFORMATION: females bear eggs from September through to March of the following year.

This species is fished with pots in UK coastal regions, sometimes with nets, locally by SCUBA diving. It is usually marketed fresh, often whole. There are small but important UK fisheries in south western Cornwall and the Isles of Scilly where it is usually fished from spring until autumn; it is more intensively fished in France and Portugal and less so in the Mediterranean.

The closely related Pink Spiny Crawfish, *P. mauritanicus* (Pl. 14) occurs from south-western Ireland to southern Senegal and in the western Mediterranean.
European vernacular names for this species are: Langouste de Mauritanie (Fr); Langosta Rosada, Langosta Mora (Sp); Aragosta, Aragosta Mauritanica (It).

This species is distinguished from *P.elephas* by having only 1-2 (rarely 3) small spines on the hind margin of the abdominal pleura (fig. 6.2, D) and the male is without a disto-external spine on the propodus (**pro**, fig. 6.2, E) of the first pereiopod.

Colour: pink to pinkish red; pereiopods banded with white to pink, abdominal segments marbled.

Fished mainly off North West Africa and sporadically in Mediterranean.

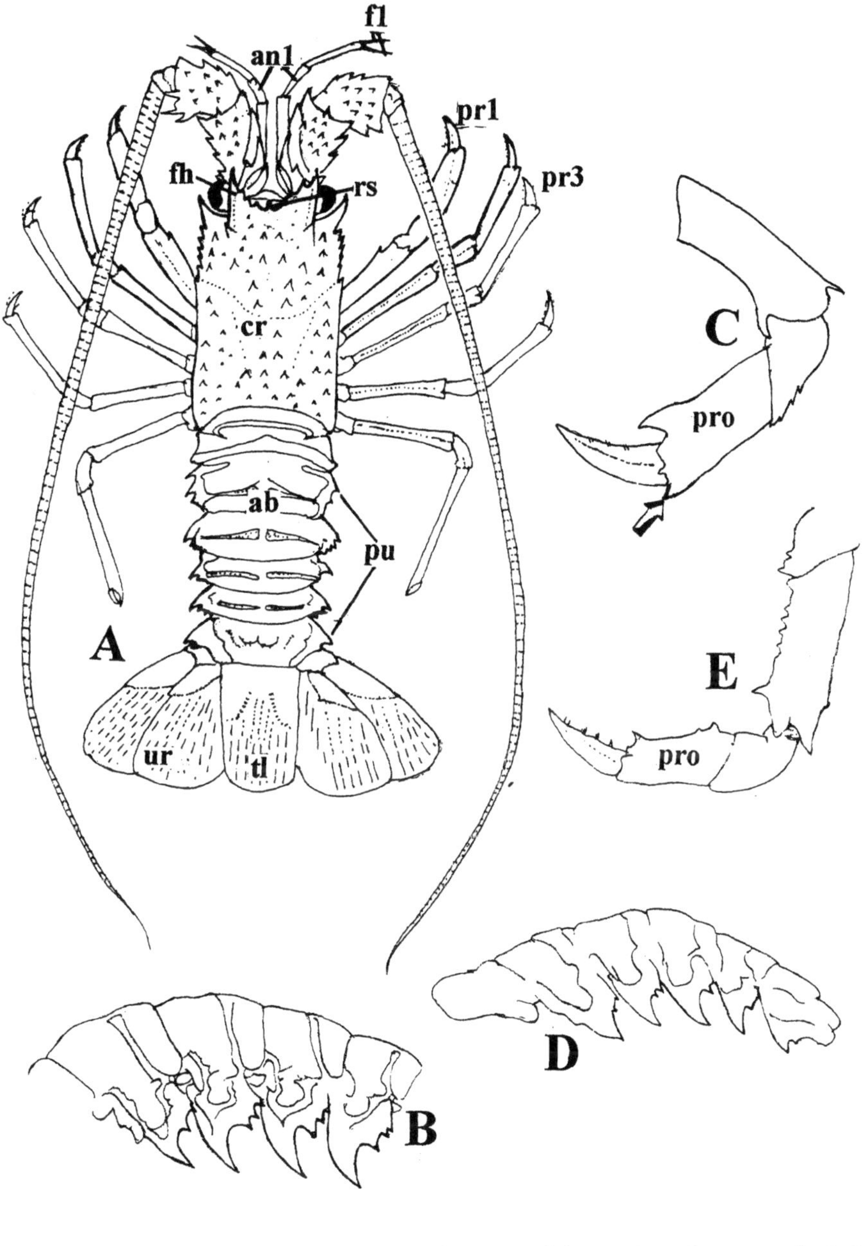

Fig. 6.2 *Palinurus elephas*, European Spiny Crawfish. A. Dorsal aspect of whole crawfish. B. Abdominal pleura of left side. C. First pereiopod from right side. *Palinurus mauritanicus*, Pink Spiny Crawfish. D. Abdominal pleura from left side. E. First pereiopod from right side. For explanation of annotations see opposite page.

Panulirus regius De Brito Capello, 1846
Fig. 6.3, Pl. 15.
COMMON SYNONYMS: *Palinurus rissonii.*
DERIVATION: Panulirus-an anagram of Palinurus (Gr)-steersman and Trojan sailor to Aenaeus; proverbially used for advisor or pilot; regius (Lt)-royal probably on account of size and colour.
VERNACULAR NAMES: Royal Spiny Crawfish, Royal Spiny Lobster (UK); Langouste Royal (Fr); Langosta Real (Sp); Bakhouche, Azeffane (Mr); Lagosta Verde (An); Soum (Ga); Blue Crawfish (Gh).
DISTINCTIVE FEATURES: abdominal segments with transverse groove (**tg**) interrupted in mid-line; antennular plate (**ap**) with four large spines arranged in two longitudinal rows; flagella (**f1**) of first antennna (**an1**) long; third maxilliped (see **mxpd**, fig 1.2) without an exopod.
COLOUR: olive green, posterior margin of each abdominal segment with a transverse white band, anteriorly and posteriorly emarginated by a brown to black band, a yellow or white spot near base of each pleuron (**pu**); tail fan uniform green to brownish in colour.
TOTAL LENGTH: 25-35 cm.
DISTRIBUTION: North Eastern Atlantic-west coast of African from Angola to Morocco and western Mediterranean.

The Royal Spiny Crawfish inhabits rocky substrates often between 5 and 15 m, but occurs down- wards to depths of 40 m.
GENERAL INFORMATION: this crawfish is fished along the African west coast and in the Mediterranean with tangle nets, trawls, pots and creels. Whole specimens are often marketed fresh in Spain and France or as frozen tails in other countries. Live specimens are occasionally seen in public aquaria and in local restaurants.

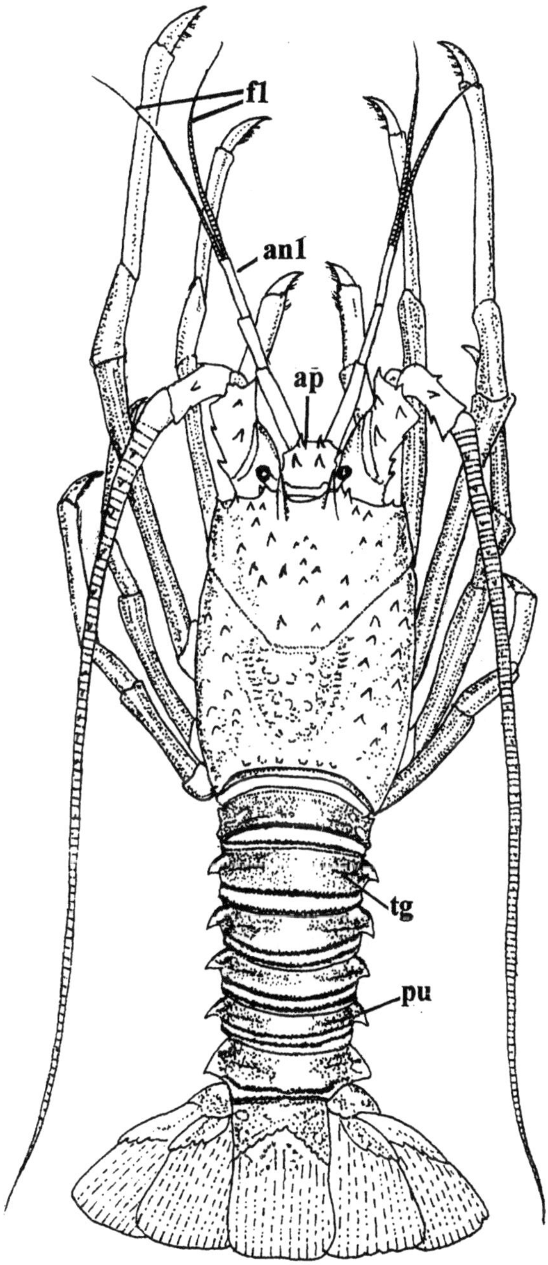

Fig. 6.3 *Panulirus regius*, Royal Spiny Crawfish. For explanation of annotations see opposite page.

Scyllarides latus (Latreille, 1802)
Fig. 6.4, Pl. 17.
COMMON SYNONYMS: none.
DERIVATION: from skyllaros (Gr)-a kind of crab, with Greek patronymic termination-ides; latus (Lt)-broad.
VERNACULAR NAMES: Mediterranean Slipper Lobster, Flat Lobster (UK); Björnkräfta (Sw); Bjørnekreps (Nw); Grosser Bärenkrebs (Gn); Beerkreeft (Nd); Grande Cigale, Grosse Cigale (Fr); Cigarra, Cigarra de Mar (Sp); Cavaco (Pt); Magnosa, Cicala Grande (It); Astakoudaki Megalo (Gr); Büyük Ayi Istakozu (Tr); Cigale Noire, Farzit (Tn).
DISTINCTIVE FEATURES: carapace (**cr**) broad, frontal region with a distinct rostrum (**rs**), eyes located near carapace margin; dorsal surface of carapace and abdomen with numerous well developed tubercles; outer two large spots on first abdominal segment (**ab1**) with the inner margins concave, posterior margin of second segment (**ab2**) with a posterior tooth (**pt**); second antenna (**an2**) developed as a flat plate without distal marginal teeth; first to third pereiopods without claws, carpus (**crp**) of first pereiopod with a dorsal groove (**dg**).
COLOUR: reddish brown with touches of violet or blue, large red spots on first segment of abdomen.
TOTAL LENGTH: maximum 45 cm.
DISTRIBUTION: North Eastern Atlantic-from coast of Portugal southwards to Cape Verde Islands. Whole of Mediterranean, absent from Black Sea.

The Mediterranean Slipper Lobster occurs on rocky and sandy substrates in depth 4-100 m.

GENERAL INFORMATION: females with eggs occur from June-August.

Fished by all methods. Fresh or frozen whole specimens are occasionally seen in Mediterranean (particularly Israel) markets or restaurants where they usually command quite high prices.

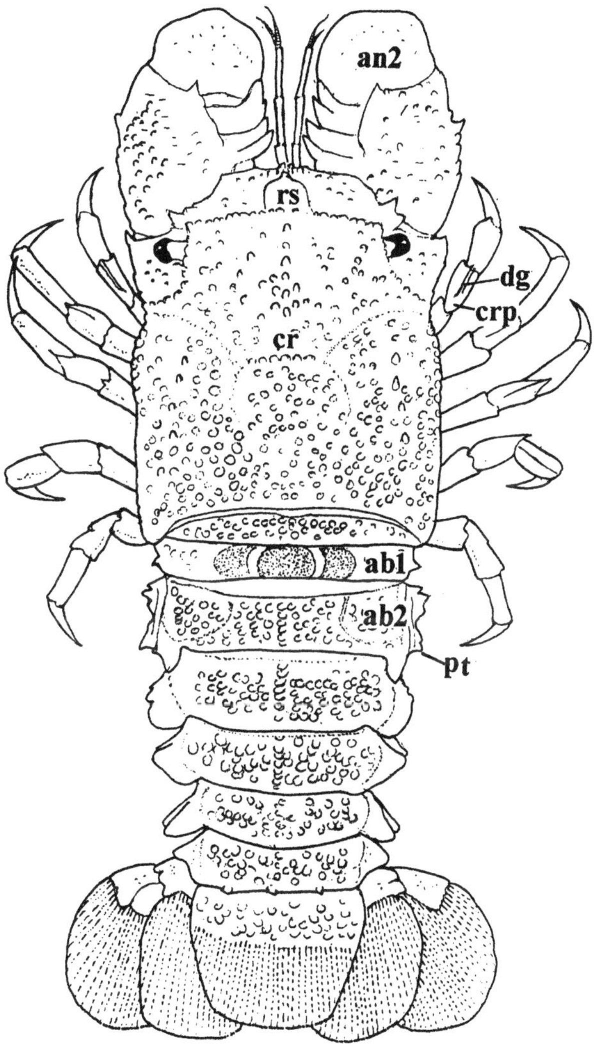

Fig. 6.4 *Scyllarides latus*, Mediterranean Slipper Lobster. For explanation of annotations see opposite page.

Scyllarus arctus (Linnaeus, 1758)
Fig. 6.5, Pl. 18.
COMMON SYNONYMS: none.
DERIVATION: from skyllaros (Gr)-a kind or crab or perhaps derived from skullo (Gr)-torn or ragged (alluding to antennal margin); artus (Lt)-narrow (ie. compared with other species).
VERNACULAR NAMES: Small European Locust Lobster, Shovel-nosed Lobster, Sand Lobster, Flat-nosed Lobster (UK); Björnkräfta (Sw); Espanjanhummeri (Fi); Bjørnekreps (Nw, Dn); Kleiner Barenkrebs (Gn); Beerkreeft (Nd); Petite Cigale (Fr); Santiaguito (Sp); Lagosta da Pedra (Pt); Cicala di Mare (It); Lyra, Astakoudaki (Gr); Küçük Ayi istakozu (Tr); Ziz el Bahr (Tn).
DISTINCTIVE FEATURES: carapace (**cr**) moderately broad, eyes (**ey**) located near carapace margins; dorsal surface of abdomen elaborately sculptured, a configuration of dorsal mid-line lobules present on second (**ab2**) to fifth segment; second antenna (**an2**) developed as a flat plate with conspicuous distal marginal lobes or teeth; first (**pr1**) to third (**pr3**) pereiopods without chelae, fifth (**pr5**) with a small claw (**cl**).
COLOUR: brownish to reddish; apices of spines and tubercles often whitish, abdomen with orange and blue spots; pereiopods banded blue.
TOTAL LENGTH: 5-10 cm rarely 12-16 cm.
DISTRIBUTION: North Eastern Atlantic-southern British Isles to Guinea, West Africa. Distributed throughout Mediterranean but apparently absent from Black Sea.

The Small European Locust Lobster occurs on muddy and rocky substrates in depths of 4-50 m.

GENERAL INFORMATION: females carrying eggs occur from February until April.

The species is fished by all methods. Fresh whole specimens are occasionally seen in Mediterranean fish markets where they sometimes commands quite high prices. The abdominal meat is considered high in quality but sparse in quantity and is often used to make soups.

Small specimens of this species are sometimes misidentified as the Pigmy Locust Lobster (*Scyllarus pygmaeus*) recognised by a short transverse hair-lined groove on the anterior parts of the second to sixth abdominal segments and only visible when the abdomen is bent.

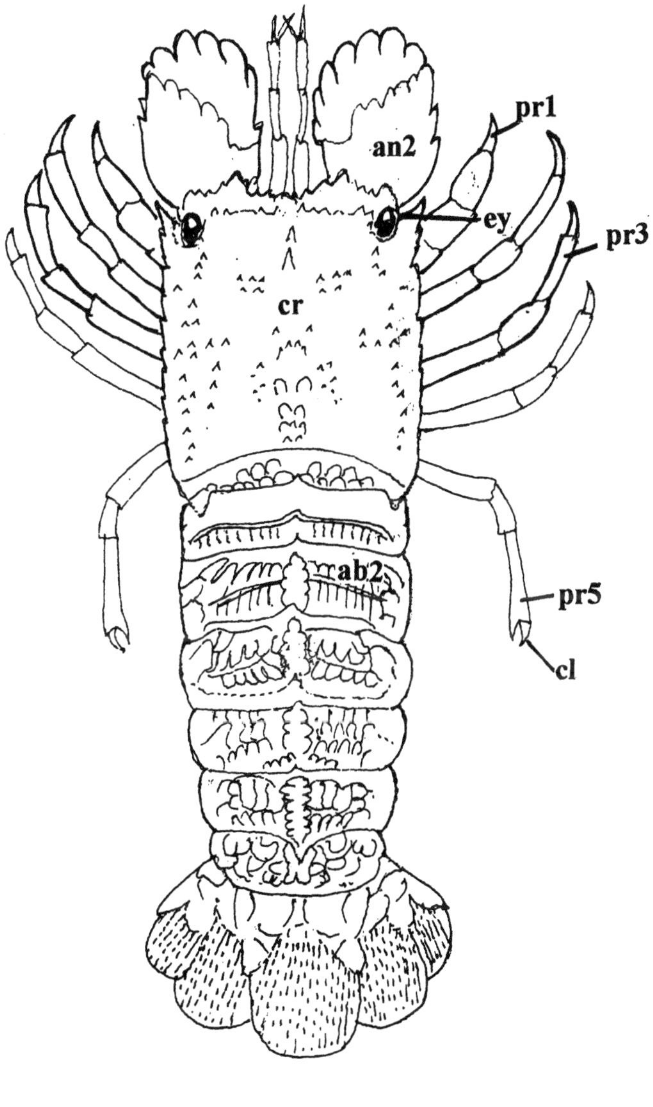

Fig. 6.5 *Scyllarus arctus,* Small European Locust Lobster. For explanation of annotations see opposite page.

FURTHER READING

Allen, J.A. (1967) *The Fauna of the Clyde Sea Area. Crustacea: Euphausiacea and Decapoda.* Scottish Marine Biological Association, Millport (pages 56 & 88-89 for crawfishes and Shovel-nosed lobsters).

Holthuis, L.B. (1991) Volume 13, Marine Lobsters of the World. FAO Species Catalogue. *FAO Fishery Synopsis,* **125**, FAO Rome (pages 90-170 for crawfishes and pages 171-228 for shovel-nosed lobsters).

Zariquiey Alvarez, R. (1968) Crustáceòs Decápodos Ibéricos. *Investigación Pesquera,* **32**, 1-510 (pages 212-222 for crawfishes and shovel-nosed lobsters).

7

Mud Lobsters (Thalassinoidea)

7.1 NATIVE SPECIES

The taxonomic status of some native species of mud lobsters is currently being revised but at least seventeen species are represented in the Northeastern Atlantic/Mediterranean regions. Only two are figured here that are considered to be the commonest. Distinctive features of mud lobster families are discussed in section 7.4.

7.2 LIFE HISTORY

The eggs of mud lobsters hatch as small shrimp-like planktonic larvae (fig. 7.1). These moult usually through five stages before developing into the benthic stage that makes burrows in the sea bed.

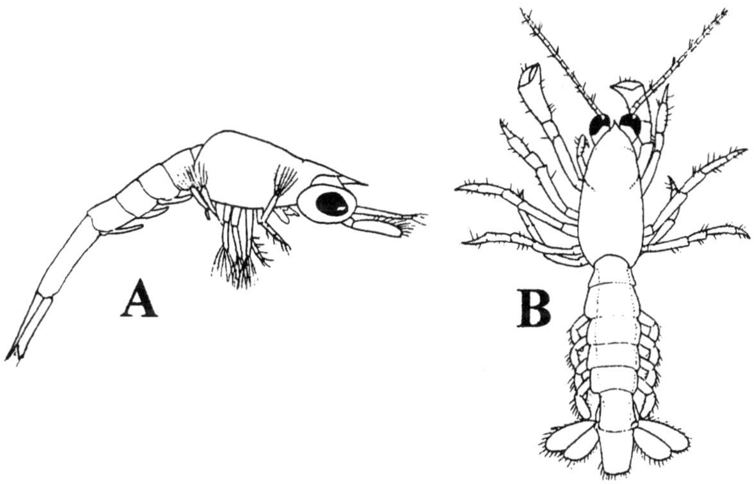

Fig. 7.1 Development of mud lobsters. A. Late free swimming planktonic larval stage, total length 3.5 mm. B. First benthic stage, total length 4.5 mm.

7.3 BIONOMICS

Very few species of mud lobsters are consumed by humans and not by European communities. The Ghost Mud Lobster and the Mediterranean Mud Lobster figured are collected for fish bait, chiefly by sport fishermen.

Mud lobsters construct elaborate branching tunnels below the sea bed. The walls of these are usually consolidated by binding sand or mud with mucus secreted by glands in the thoracic limbs. This mixture is then plastered against the tunnel wall along with pieces of algae, small stones, or other material.

Some mud lobsters, for example species of *Upogebia*, often occur in pairs. Other animals, such as bivalve molluscs, may also inhabit the tunnels.

Many species of mud lobsters are found only in offshore waters. In the southern regions of the British Isles the distribution of a few species extend into suitable muddy regions on the lowest levels of sheltered shores exposed only at spring tides. In more southern locations of Europe and in the Mediterranean the burrows of some species are sometimes obvious in lagoon-like regions sheltered from strong currents.

In captivity mud lobsters emerge to forage for food on the substrate surface but in the sea nearly all their entire life is probably spent within the confines of the burrows. The beating pleopods create water currents that draw in suspended particles that are sieved by the mouthparts for food. Mud lobsters apparently can subsist on a diet composed almost entirely of diatoms [marine algae] although they will attack and consume other creatures, such as ragworms, that enter their tunnels.

Mud lobsters are sometimes present in large numbers in offshore waters and their tunnels (fig. 7.2) can extend in excess of one metre in depth and over wide areas, even in clay. They are only occasionally captured when trawls disturb the sea bed to a sufficient depth to evict them from burrows.

7.4 IDENTIFICATION

The native species of mud lobsters belong to four families. Those assigned to the family Axiidae are without a longitudinal suture on the carapace; this is present (**ls**, fig. 3.1, E) in all species belonging to the other families. Members of the family Laomediidae all have a well defined transverse suture on both exopod (outer branch) and endopod (inner branch) of the uropods. For the remaining two families these sutures are absent, or at most an incomplete one is present only on the exopod. Species of the family Callianassidae are either without a rostrum or a very small one is present (**rs**, fig. 7.3, A) whereas the rostrum is always well developed (**rs**, fig. 7.4, A) in species belonging to the family Upogebiidae.

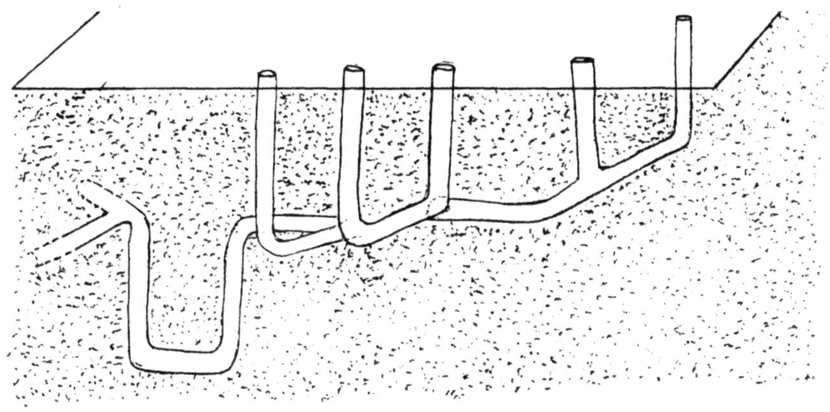

Fig. 7.2 Diagram of interconnecting tunnels constructed by mud lobsters.

Callianassa tyrrhena (Petagna, 1792)
Fig. 7.3.
COMMON SYNONYMS: *Callianassa subterranea.*
DERIVATION; kalos (Gr)-beautiful; nassa (Lt)-narrow-necked fish basket-probably alluding to basket-like feeding appendages; tyrrhena-alluding to Tyrrhenian Sea in Mediterranean where species was first collected.
VERNACULAR NAMES: Ghost Mud Lobster, Sand Ghost Lobster/Shrimp, Mud Shrimp (UK); Maulwurfkrebs, Sandkrebs (Gn); Topo de mar (Sp); Corbola selvatica falsa, Corbols salvadega (It).
DISTINCTIVE FEATURES: body relatively soft (ie not strongly calcified); rostrum (**rs,** fig. A) very small; third maxilliped (**mxp3**) expanded to form an operculum; chela (**ce**) of right first pereiopod (**pr1**) very large, lower margin of merus (**mrs,** fig. B) with a blunt lobe; uropods (**ur,** figs A, C) without a transverse suture and slightly longer than telson (**tl,** fig. A, C), telson without spines, rounded posteriorly, about as broad as long and with posterior margin rounded.
COLOUR: whitish to translucent greyish pink.
TOTAL LENGTH: between 6-7 cm.
DISTRIBUTION: North Eastern Atlantic-from southern Norway and Sweden to West coast of Africa as far as Mauritania. Distributed throughout the entire Mediterranean but perhaps not in the Black Sea.

The Ghost Mud Lobster occurs in muddy substrates (into which it makes tunnels). It rarely inhabits the lowest shore but is more commonly found in offshore substrates covered by water of 2-3 m depth, but may extend downwards into water depths of about 60 m.
GENERAL INFORMATION: this mud lobster constructs tunnels of some 40 cm in depth. It is very occasionally trawled by fishing boats and used for fish bait by anglers when captured.

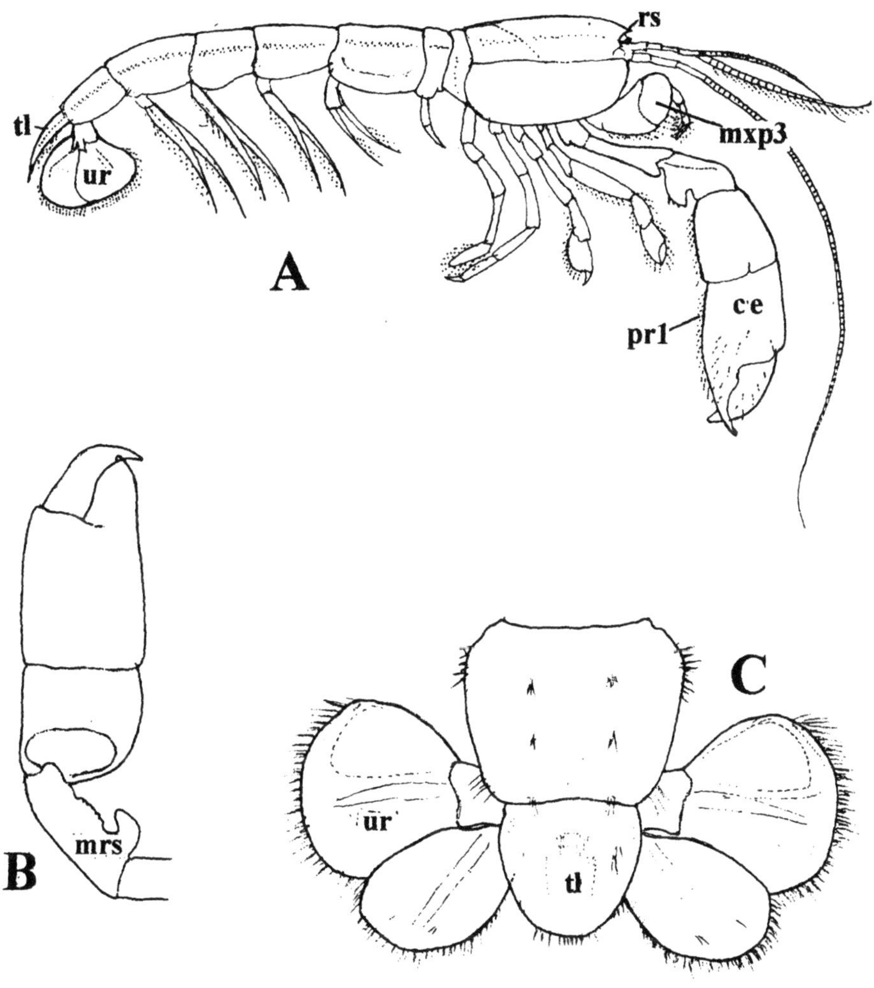

Fig. 7.3 *Callianassa tyrrhena*, Ghost Mud Lobster. A. Lateral right aspect of whole shrimp. B. Right first pereiopod. C. Sixth abdominal segment, telson and uropods from dorsal aspect. For explanation of annotations see opposite page.

Upogebia pusilla (Petagna, 1792)
Fig. 7.4.
COMMON SYNONYMS: *Thalassina littoralis*; *Upogebia littoralis*.
DERIVATION: upo (Ni)-under or below; ge (Gr)-pertaining to earth; bia, derived from biota (Ni)-pertaining to habits of the crustacean;. pusillus (Lt)-very small, weak.
VERNACULAR NAMES: Mediterranean Mud Lobster, Mud Shrimp (UK); Maulwurfkrebs (Gn); Crevette Fouisseuse (Fr); Grillo Real Marino (Sp); Corbola, Cicaledda (It); Mamun (Tr).
DISTINCTIVE FEATURES: rostrum (**rs,** fig. A) well developed; antero-lateral margins of carapace (**al**) with small tooth; chelae (**ce**) of first pair of pereiopods (**pr1**) more or less equal in size, largest noticeably distally broadened in male; telson (**tl,** fig. B) broad and subtruncate, uropods (**ur**) slightly shorter than telson.
COLOUR: whitish to translucent yellow.
TOTAL LENGTH: maximum 6.5 cm.
DISTRIBUTION: North Eastern Atlantic-from Brittany to Mauritania, West Africa. Occurs throughout whole of the Mediterranean region, including the Black Sea.

The Mediterranean Mud Lobster inhabits muddy or muddy sand substrates (in which it makes tunnels) and occurs from the lowest part of the shore downwards to depths of about 45 m.

GENERAL INFORMATION: tunnels made by this lobster are often of a U-or Y-shape construction.

Females bearing eggs occur during June and August in the Mediterranean region.

This lobster is collected for bait (particularly in the Mediterranean) by sport fishermen.

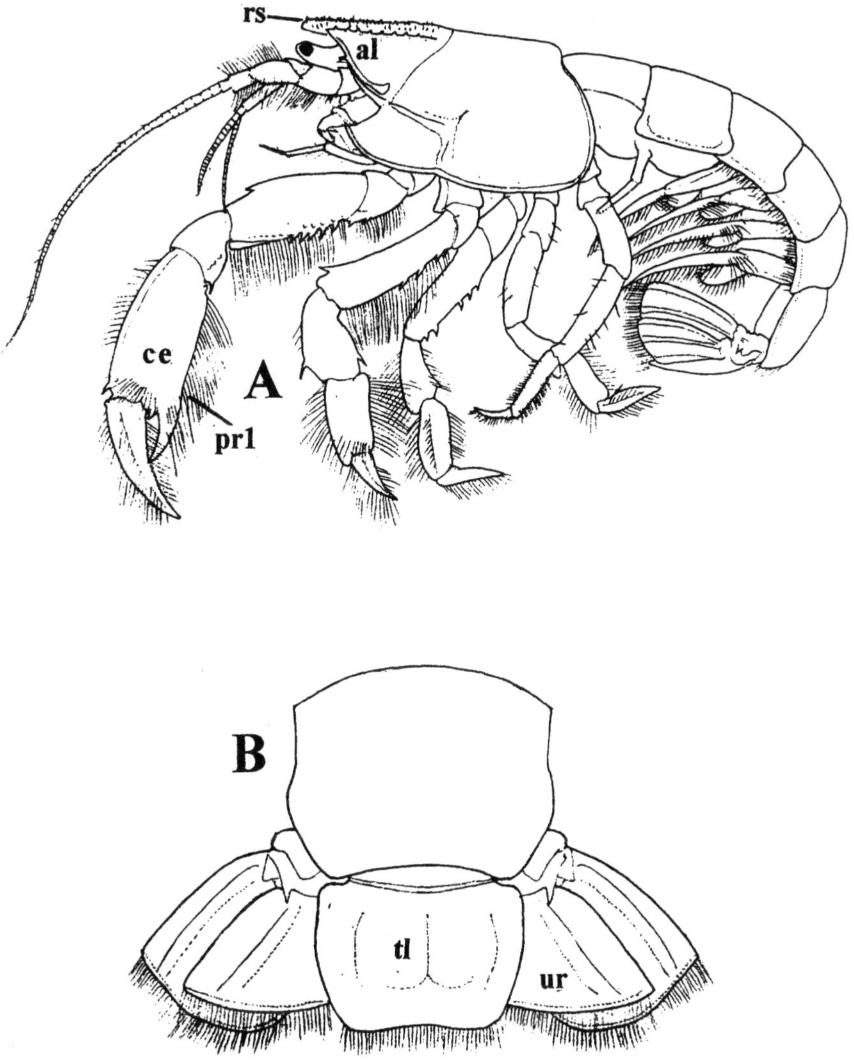

Fig. 7.4 *Upogebia pusilla*, Mediterranean Mud Lobster. A. Left lateral aspect of whole lobster B. Sixth abdominal segment, telson and uropods from dorsal aspect. For explanation of annotations see opposite page.

FURTHER READING

Allen, J.A. (1967) *The Fauna of the Clyde Sea Area. Crustacea: Euphausiacea and Decapoda.* Scottish Marine Biological Association, Millport (pages 57 & 90 for mud lobsters).

Atkinson, R.J.A. and Taylor, A.C. (1988) Physiological ecology of burrowing Decapods, in Fincham, A.A. and Rainbow, P.S. (eds). Aspects of Decapod crustacean biology. *Symposium of the Zoological Society of London*, **59**, 201-226.

Holthuis, L.B. (1991) Volume 13, Marine Lobsters of the World. FAO Species Catalogue. *FAO Fishery Synopsis*, **125**, FAO Rome (pages 229-253 for mud lobsters).

Nash, R.D.M., Chapman, C,J., Atkinson, R.J.A and Morgan, P.J. (1984) Observations on the burrows and burrowing behaviour of *Calocaris macandreae* (Crustacea: Decapoda: Thalassinoidea). *Journal of Zoology, London*, **202**, 425-439.

Zariquiey Alvarez, R. (1968) Crustáceòs Decápodos Ibéricos. *Investigación Pesquera,* **32**, 1-510 (pages 223-232 for mud lobsters).

8

Hermit and Stone Crabs (Paguroidea)

Hermit crabs described in this book belong to the families Diogenidae and Paguridae. Their abdomens are soft, spirally curled, and usually concealed within empty gastropod shells held in place by the uropods (**ur**, fig. 1.3, A), braced against the shell's internal wall. By comparison, the abdomen of Stone Crabs (family Lithodidae) is loosely held beneath the cephalothorax. It is also soft but bears calcified plates (fig. 8.7, C, D); these are conspicuously asymmetrically arranged in the female. The pleopods and uropods are absent.

8.1 NATIVE SPECIES OF HERMIT CRABS

Forty six species of hermit crabs inhabit the North Eastern Atlantic and Mediterranean regions, none are fished for human consumption. The commonest species occurring on middle to low shores of North Eastern Atlantic countries are small specimens of Bernard's Hermit Crab, *Pagurus berhardus*. Considerable populations of large crabs of this species and also of Prideaux's Hermit Crab, *Pagurus prideaux* commonly live in offshore waters. On Mediterranean shores Roux's Hermit Crab, *Diogenes pugilator* and Latreille's Hermit Crab, *Clibanarius erythropus* frequently occur.

8.2 LIFE HISTORY OF HERMIT CRABS

Typical mating behaviour is described as follows. Prior to mating the male carries the female for a few hours to several days, holding her shell margin with his chelae. Females receptive to mating may probably produce a chemical [pheromone] that attracts males. A pre-mating ritual occurs in which both sexes face each other and tap shells after which they momentarily leave their shells and mate. Females already carrying fertilized eggs will also participate in mating and these eggs usually hatch soon after a further mating. The male hermit crab typically deposits sperm packets [spermatophores] on or near the coxal region of the female's third and fourth pereiopods. The spermatozoa are liberated when the female spawns thus fertilizing the eggs. These are incubated attached to oosetae on the female's small pleopods. The small shrimp-like larvae [zoeae] that hatch pass through three stages all similar (fig. 8.1, A) before changing into a form [megalopa or glaucothoe] suited for life on the sea bed (fig. 8.1, B). During the following moults to juvenile stages the abdomen assumes the characteristic coiled-shape and these young hermit crabs must find gastropod shells of suitable sizes in which to live and assume their specialised life style otherwise they will not survive.

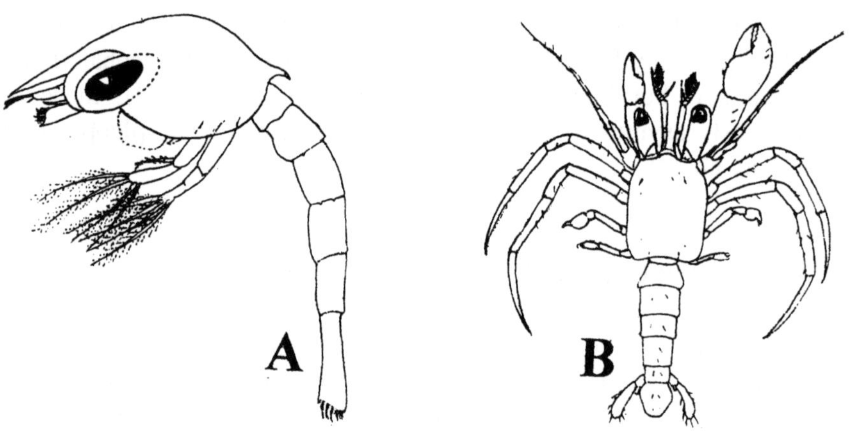

Fig. 8.1 Hermit crab development. A. First larval stage [zoea], 4 mm total length. B. First benthic stage [megalopa or glaucothoa], 5 mm total length.

8.3 BIONOMICS OF HERMIT CRABS

The shells occupied by hermit crabs may carry a variety of epi-fauna and flora. Common encrustations are barnacles, worms tubes [annelids], sponges and sea firs [hydroids]. Large whelk shells inhabited by Bernard's Hermit crabs living offshore are sometimes covered with Couch's Anemone, *Calliactes parasitica* and occasionally the Ragworm, *Nereis furcata* occurs within the shell. Prideaux's Hermit Crab (p. 88) is sometimes present among fish catches landed at quaysides. This species is usually invested with one or more Cloak Anemones, *Adamsia carciniopados* [= *A. palliata*] whose bases fold around and usually totally conceal the gastropod shell.

Hermits crabs need to change to larger shells in order to grow. The availability of empty suitable shells of larger sizes can affect progressive growth and fighting will occur among crabs to acquire or defend shells. An exception is Prideaux's Hermit Crab since sufficient shelter is usually provided by the base of the Cloak Anemone as it grows and gradually dissolves the shell, thus allowing the Crab to increase in size without need to find another shell of suitable dimensions.

Some regions in offshore waters contain considerable populations of hermit crabs but their rôle in the ecology of seabed communities is little understood. They provide a valuable food source for bottom feeding fishes and are, no doubt, effective scavengers.

Hermit crabs are often present in large numbers on shores strewn with gastropod shells; rapidly moving shells is a sure indication they are occupied by hermit crabs.

8.4 IDENTIFICATION OF HERMIT CRABS

Hermit crabs are less easy to identify than many other crab-like crustaceans. Two obvious featured can be used to assign specimens to one or the other of the two families included in this book. The few species belonging to the family Diogeniidae have the basal parts of the third pair of maxillipeds (**mxpd 3**, fig. 1.3, B) placed close together and the chelipeds are almost equal in size or the left is larger than the right. The numerous species belonging to the family Paguridae have an appreciable gap between the pair of third maxillipeds and the right cheliped is usually larger than the left. For further information about the taxonomy of North eastern Atlantic species the reader should consult the references in 'Further Reading' at the end of this chapter.

Because of the soft nature of their bodies, hermits crabs are not particularly easy to measure. The most satisfactory measurement is the length of the hardest anterior portion of the carapace termed the shield (labelled **sh** in the figures).

Pagurus bernhardus (Linnaeus, 1758)
Fig. 8.2.
COMMON SYNONYMS: *Eupagurus bernhardus.*
DERIVATION: pagurous (Gr)-a crab; berhardus-probably named after St Bernard of Clairvaux founder of a Cistercian order of monks.
VERNACULAR NAMES: Bernard's Hermit Crab, Common Hermit Crab, Soldier Crab (UK); Eremitkräfta (Sw); Erakkorapu (Fi); Eremittkreps (Nw); Eremitkrebs (Dn); Heremietkrab (Nd); Nordsee-Einsiedlerkrebs (Gn); Bernard l'Hermite (Fr); Bruja (Sp); Casa-alugada (Pt).
DISTINCTIVE FEATURES: permanent inhabitant of marine snail shells (gastropods). Body (fig. 8.2, A) relatively soft; abdomen (**ab**) appearing spirally curled when crab is removed from shell; a well developed rostrum (**rs**) present; right chelae (**ce**) of first pereiopod (**pr1,** fig. A) larger than left; width of propodus (**pro**) of both left (fig. B) and right (fig. C) chelae slightly exceeding half length, dorsal surface of propodus slightly convex and with oval tubercles; propodus of right chela often with discernible rows of larger tubercles; dactylus (**dcy**, fig. D-only distal part shown) of second (**pr2**) and third (**pr3**) pereiopods with short spines arranged close together.
COLOUR: variable, reddish brown to violet; proximal segments of second and third pereiopods often with reddish brown transverse bands, tubercles on chelae white.
SHIELD LENGTH: maximum about 2 cm.
DISTRIBUTION: North Eastern Atlantic-from Spitzbergen to coast of Portugal; also northeastwards to Murman Sea, Russia. Absent from Mediterranean.

Bernard's Hermit Crab often occurs intertidally and downwards to depths of 1800 m. Small crabs frequently inhabit rockpools whereas large crabs occur offshore on rock, stone, sand and sandy mud substrates, sometimes in considerable numbers.
GENERAL INFORMATION: females carrying eggs occur throughout the year but British inshore populations tend to breed from November to May.

This hermit crab is not considered to be of any commercial value although at one time the abdomens were sold in french markets. It is occasionally used as bait by cod and bass fishermen.

Vernacular names used in the Mediterranean for this species probably refer to Prideaux's Hermit Crab (p. 88) or to other species.

Bernard's Hermit Crab is often seen in public aquaria.

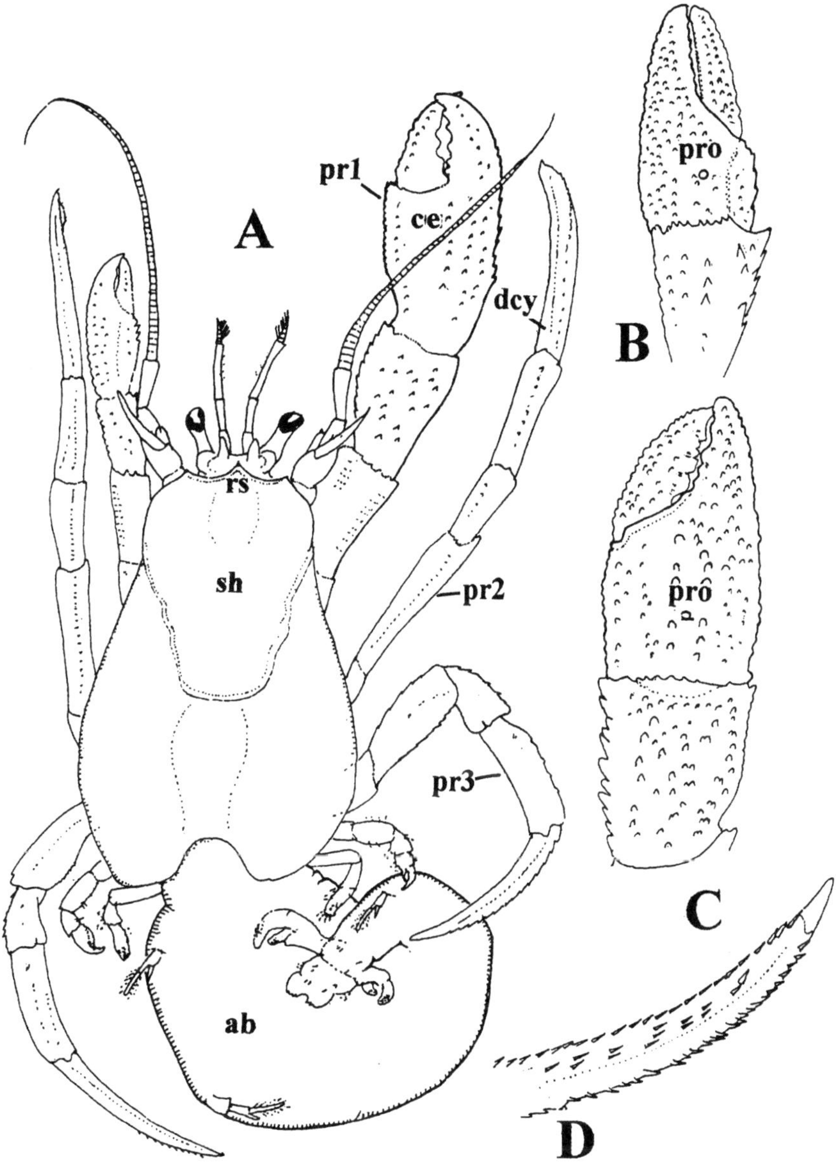

Fig. 8.2 *Pagurus bernhardus*, Bernard's Hermit Crab. A. Dorsal aspect of whole crab removed from gastropod shell. B & C. Left and right first pereiopods (chelae) respectively. D. Distal portion of second pereiopod dactyl from inner aspect. For explanation of annotations see opposite page.

Pagurus prideaux Leach, 1815
Fig. 8.3.
COMMON SYNONYMS: *Pagurus solitarius; Eupagurus prideauxi.*
DERIVATION: pagouros (Gr)-a crab; prideaux after the Devonshire naturalist Charles Prideaux of Kingsbridge.
VERNACULAR NAMES: Prideaux's Hermit Crab (UK); ?Paguro Bernado (It); ?Skaltsini (Gr); ?Kesis Istakozu (Tr). These Italian, Greek and Turkish names are used for Bernard's Hermit Crab but probably should be applied to Prideaux's Hermit Crab since the former species has not been reported from the Mediterranean.
DISTINCTIVE FEATURES: gastropod shell inhabited by this crab are often completely enveloped by a Cloak Anemone. Abdomen spirally curved when crab is removed from shell; shield (**sh**) broad, rostrum absent; right chela (fig. C) larger than left (fig. B); width of propodus (**pro**) of chelae of larger cheliped equal to or less than half length, outer propodal surface of both chelipeds convex, with numerous small tubercles, none arranged in a distinct pattern; dactylus (D-distal part shown only) of second and third pereiopods relatively thin and with raised longitudinal setosed ridge.
COLOUR: reddish brown, carapace sometimes speckled with white, second and third pereiopods often with transverse violet-black bands. Eyestalks often with red band.
SHIELD LENGTH (**sh**): maximum about 1.5 cm.
DISTRIBUTION: North Eastern Atlantic-from Bergen (Norway) to coast of Morocco, North West Africa). Distributed throughout whole of Mediterranean including Black Sea.

Prideaux's Hermit Crab occurs chiefly offshore in depths between 7-250 m on sandy and muddy sand substrates and sometimes in large numbers.
GENERAL INFORMATION: in the North Eastern Atlantic females carrying eggs occur from January until September and in the Mediterranean from March until December.

This crab is often seen in public aquaria.

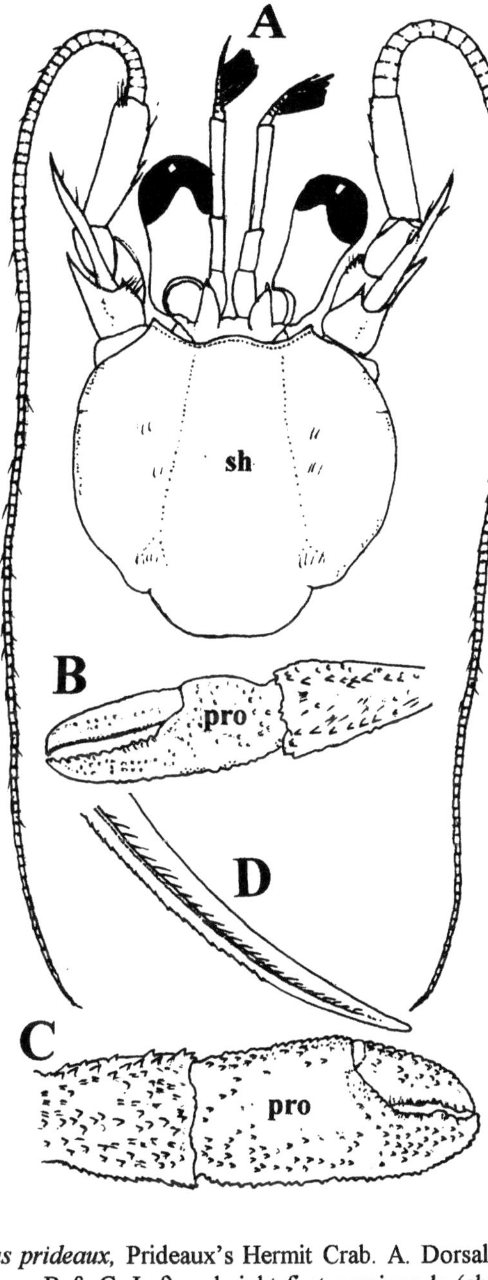

Fig. 8.3 *Pagurus prideaux,* Prideaux's Hermit Crab. A. Dorsal aspect of shield and anterior appendages. B & C. Left and right first pereiopods (chelae) respectively. D. Distal portion of second pereiopod dactyl from inner aspect. For explanation of annotations see opposite page.

Diogenes pugilator (Roux, 1829)
Fig. 8.4.
COMMON SYNONYMS: none.
DERIVATION: Diogenes (Gr)-divine originating from Zeus or perhaps alluding to the celebrated cynic who resided in a tub; pugilator (Lt)-a message bearer.
VERNACULAR NAMES: Roux's Hermit Crab (UK).
DISTINCTIVE FEATURES: abdomen spirally curved when crab is removed from shell. A small longitudinal process (**lp,** fig. A) between basal parts of eyestalks; ocular acicles (**oc**) broad and with serrate margins; antennal flagellum (**fl**) with numerous long setae; chela of first right pereiopod (B) smaller than left, outer propodal (**pro**) surface broad and with evenly spaced tubercles.
COLOUR: carapace and limbs reddish brown, tips of chelipeds white.
SHIELD LENGTH: maximum about 4.4 mm.
DISTRIBUTION: North Eastern Atlantic-from Swansea (Wales) southwards to Angola (West Africa). Distributed throughout the whole Mediterranean including the Black Sea; also occurs in the Red Sea and Singapore.

Roux's Hermit Crab inhabits the low shore downwards to depths of about 1800 m.
GENERAL INFORMATION: females carrying eggs occur from June to July in south west England and during September in the Mediterranean. Roux's Hermit Crab is occasionally seen in public aquaria.

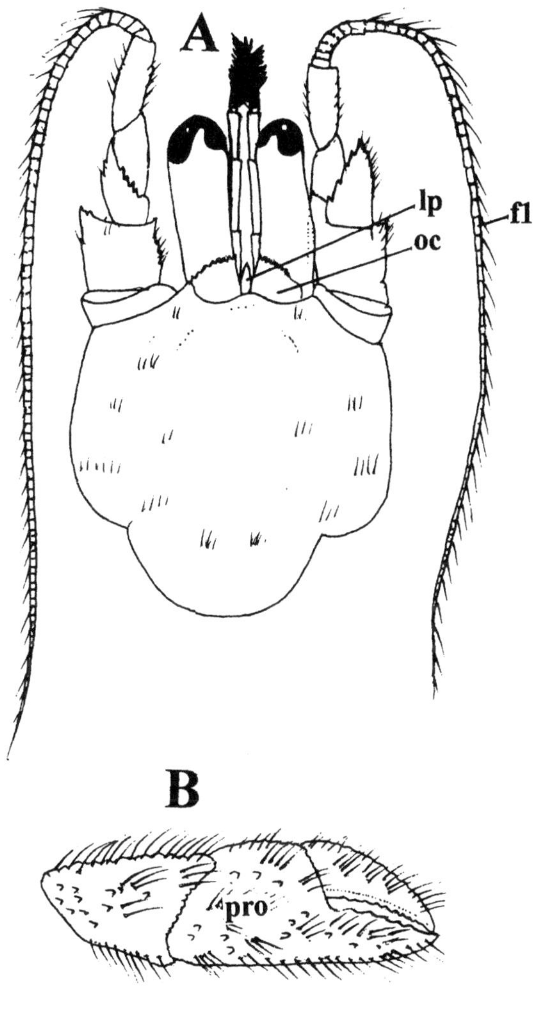

Fig. 8.4 *Diogenes pugilator,* Roux's Hermit Crab. A. Dorsal aspect of shield and anterior appendages. B. Chela of right first pereiopod. For explanations of annotations see opposite page.

Clibanarius erythropus (Latreille, 1818)
Fig. 8.5.
COMMON SYNONYMS: *Clibanarius misanthropus.*
DERIVATION: perhaps derived from klibanos (Gr)-a pot alluding to shell inhabited by crab; erythranos (Gr)-red, ruddy-alluding to colour of body and claws.
VERNACULAR NAMES: Latreille's Hermit crab (UK); Piada, Pada (Sp).
DISTINCTIVE FEATURES: abdomen spirally curved when removed from shell. Rostrum (**rs,** fig. A) well developed; eyestalks (**ey**) long; ocular acicles (**oc**) curved and with serrated margins; chelipeds (right shown in fig. B), equal in size, with long setae and corneous apices.
COLOUR: body reddish, limbs red-banded, corneous tips of chelae deep black.
MAXIMUM SHIELD LENGTH: 8 mm.
DISTRIBUTION: North Eastern Atlantic-from North Cornwall (England) to Canary Islands and whole of Mediterranean, including Black Sea.
Latreille's Hermit Crab occurs intertidal downwards to 40 m depth.
GENERAL INFORMATION: females carrying eggs occur from May until August in Mediterranean.

This crab is occasionally seen in public aquaria.

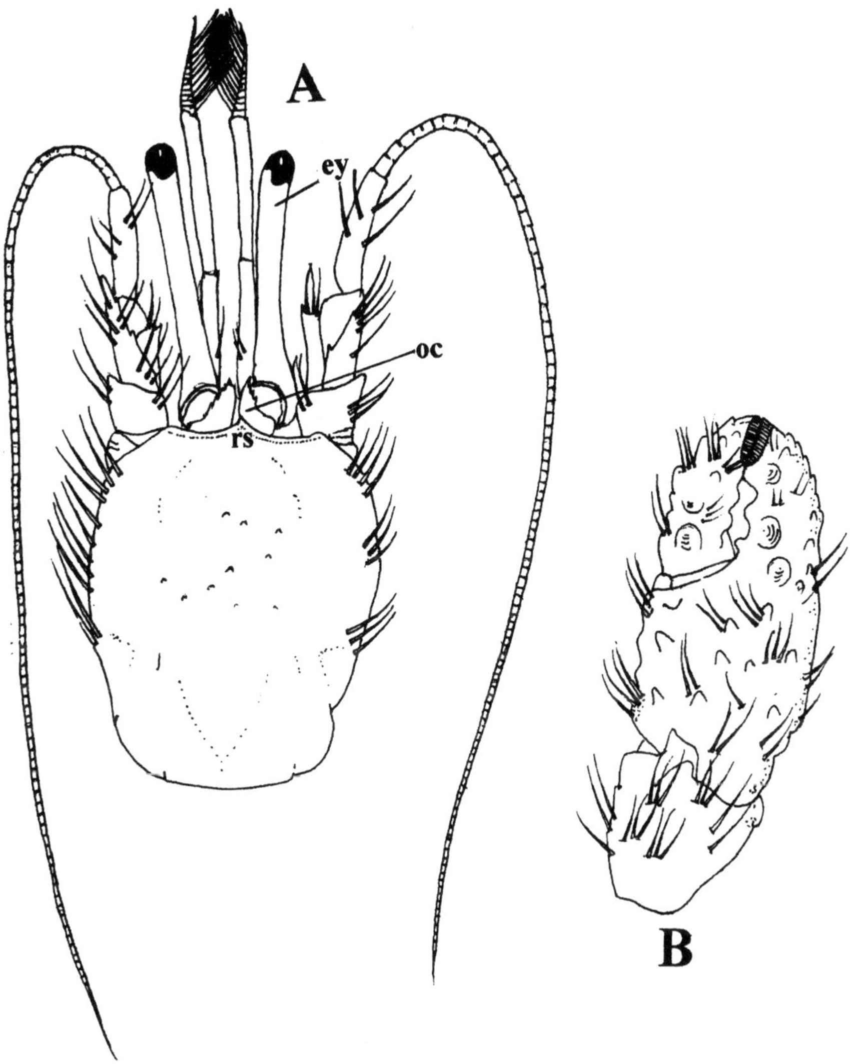

Fig. 8.5 *Clibanarius erythropus,* Latreille's Hermit Crab. A. Dorsal aspect of shield and anterior appendages. B. Chela of right first pereiopod. For explanations of annotations see opposite page.

8.5 NATIVE SPECIES OF STONE CRABS

Five species of stone [lithodid] crabs are known from the North Eastern Atlantic region, none occur in the Mediterranean. The Northern Stone Crab (*Lithodes maja*) is sometimes taken by fishermen trawling in moderate to deep offshore waters. The rare deep water Grimaldi's Stone Crab (*Neolithodes grimaldi*) is seldom caught. Three species have no vernacular names. These are *Paralomis bouvieri*, *P. microps* and *P. spectabilis*. They are uncommon and known only from a few specimens.

8.6 LIFE HISTORY OF STONE CRABS

Mating of native species has not been described. In common with other crabs the eggs, that are large, are incubated attached to the pleopod setae.

The larvae [zoeae] hatch as a shrimp-like form (fig. 8.6, A) and closely resemble larvae of hermit crabs except that the telson (B) has a broad and deep median cleft. The larvae moult through only one additional planktonic stage before changing into a crab-shaped (C) megalopal stage [sometimes termed glaucothoe]; this adopts a benthic life on mud or sandy mud of the sea bed.

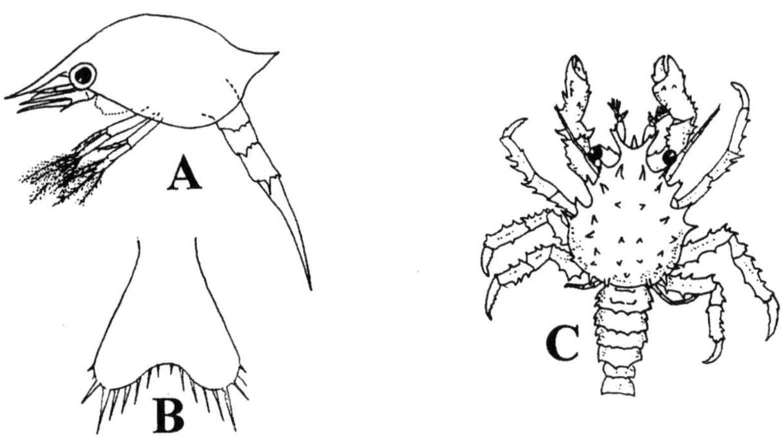

Fig. 8.6 Development of the Northern Stone Crab. A. First stage larva [zoea], total length 8.5 mm. B. Telson of same showing median cleft. C. First benthic stage [megalopa or glaucothoa], total length 4.3 mm.

8.7 COMMERCIAL EXPLOITATION OF STONE CRABS

There are in excess of 80 known species of stone crabs and some that occur in other regions are commercially fished, these are often termed 'King Crabs' (chapter 11). Although moderate numbers of our native Northern Stone Crab are occasionally caught by vessels trawling for fish it is rarely marketed.

8.8 IDENTIFICATION OF STONE CRABS

The Northern Stone Crab has the apex of the rostrum (fig. 8.7, B) prominently bifurcated. All the other species mentioned have a trifurcate rostral apex. Grimaldi's Stone Crab has very long thin spines on carapace and limbs; these are short in the species of *Paralomis*.
Paralomis bouvieri has the entire dorsal surface of the carapace covered with short spines, only a few are present in *P. spectabilis,* this region is granular in *P. microps*.

The Northern Stone Crab is occasionally misidentified as the Thorn-Back Spider Crab (*Maja squinado*)-a 'true' brachyuran crab and in which five pairs of pereiopods are always well developed and visible compared with the four pairs of stone crabs.

Lithodes maja (Linnaeus, 1758)
Fig. 8.7.
COMMON SYNONYMS: *Lithodes maia.*
DERIVATION: lithos (Gr)-a stone; maia (Gr)-a large kind of crab.
VERNACULAR NAMES: Northern Stone Crab, Norway Crab, Great Norway Crab, Devil Crab, Prickly Crab, (UK) Fanse Krabber, Trolkrabber (Nw); Augustinus Krab, Gaddakrabba, (Sc); Steinkrabbe (Gn); Crabe de Roche, Crabbe Epineux, Crabe Lithodid (Fr).
DISTINCTIVE FEATURES: carapace (**cr**, fig. A) and limbs with conspicuous spines, usually only four pairs of limbs visible, fifth pair concealed beneath carapace posterior margin; rostrum (fig. B) terminally bifid; abdomen (viewed ventrally) composed of distinct chitinous plates; these are asymmetrically arranged in female (fig. C) but more symmetrical in male (fig. D).
COLOUR: body and limbs reddish-brown, spines darker; limbs sometimes with light brown bandings.
CARAPACE LENGTH: maximum 14.5 cm.
DISTRIBUTION: North Eastern Atlantic-from Greenland southward probably as far as Netherlands; also Canadian/American Atlantic southward to New Jersey. Absent from Mediterranean.

The Northern Stone Crab occurs in depths from 65-790 m on muddy to shingle substrates.
GENERAL INFORMATION: female crabs with eggs occur from December through to April of the following year.

This crab is occasionally landed by fishermen, particularly in northern ports, and sometimes misidentified as the Thornback Spider Crab (p. 172).

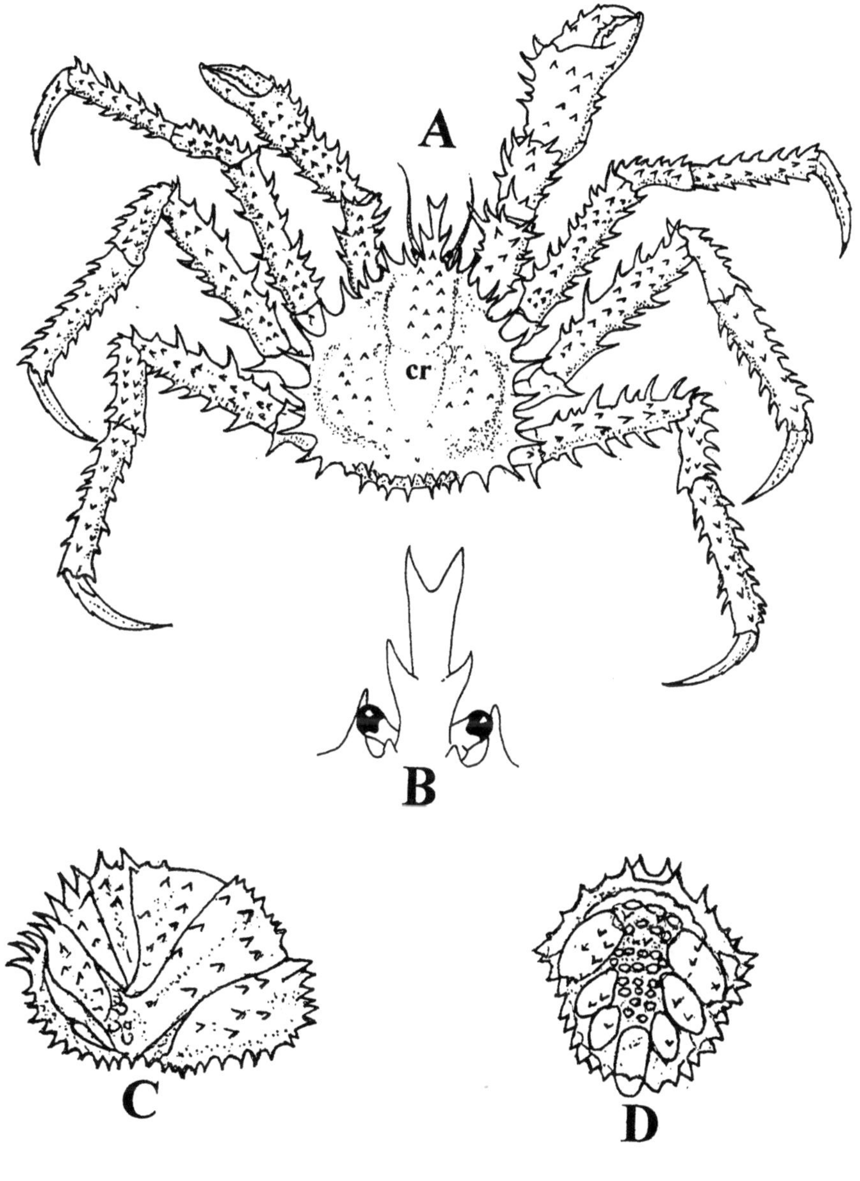

Fig. 8.7 *Lithodes maja*, Northern Stone Crab. A dorsal aspect of whole crab. B. Rostrum. C. Abdomen of female. D. Abdomen of male, both shown from ventral aspect. For explanation of annotation see opposite page.

FURTHER READING

Allen, J.A. (1967) *The fauna of the Clyde Sea Area. Crustacea: Euphausiacea and Decapoda.* Scottish Marine Biological Association, Millport (pages 60-61 & 91-93 for stone and hermit crabs).

Crothers, J.and Crothers, M. (1983) A key to the crabs and crab-like animals of British inshore waters. *Field Studies*, **5**, 753-806 (pages 763-765 and 788-789 for hermit crabs).

Dawson, E.W. (1989) *King crabs of the World: a comprehensive bibliography.* Miscellaneous Publications of the New Zealand Oceanographical Institute, Wellington.

Dawson, E.W. and Yaldwyn, J.C. (1985) King crabs of the World or the World of King crabs: an overview of identity and distribution-with illustrated diagnostic key to the genera of Lithodidae and species of Lithodes. *Proceedings of the International King Crab Symposium*, Anchorage, Alaska, pp 69-106.

Ingle, R.W. (1993) Hermit crabs of the Northeastern Atlantic Ocean and Mediterranean Sea, Identification Guide **4**. Chapman & Hall, London.

Lancaster, I. (1988) *Pagurus bernhardus* (L)-an introduction to the natural history of hermit crabs. *Field Studies*, **7**, 189-238.

Macpherson, E. (1988) Revision of the family Lithodidae Samouelle, 1819 (Crustacea, Decapoda, Anomura) in the Atlantic Ocean. *Monografias de Zoologia Marina,* **II**, 1-153, *Instituto de Ciencias del Mar,* Barcelona.

Zariquiey Alvarez, R. (1968) Crustáceòs Decápodos Ibéricos. *Investigación Pesquera,* **32**, 1-510, (pages 233-260 for hermit crabs).

9

Squat Lobsters and Porcelain Crabs (Galatheoidea)

The abdomen of squat lobsters and porcelain crabs has symmetrically arranged segments. It is held loosely folded beneath the cephalothorax and cannot be easily fully extended. Native squat lobsters include all species belonging to the families Galatheidae and Chirostylidae. In these the body is lobster-shaped, the carapace usually quite pliable, the rostrum (**rs**, fig. 9.4) well developed, the flagellum (**fl**) of the second pair of antennae long and thin, the chelae (**ce**) of the first pair of pereiopods (**pr1**) elongated and the abdomen is folded upon itself and, at most, only partly tucked beneath the cephalothorax (figs 9.2, 9.3, A).

All porcelain crabs belong to the family Porcellanidae. The crab-shaped body is hard (well calcified), the rostrum when developed is very small, the flagellum of the second pair of antennae relatively shorter than in squat lobsters, the chelae (**ce**, fig. 9.6) of the first pair of pereiopods usually broad, flattened and not noticeably elongate; the broad abdomen is loosely bent well beneath the cephalothorax.

9.1 NATIVE SPECIES OF SQUAT LOBSTERS

At least twenty four species of squat lobsters [false lobsters] are known from waters of the North Eastern Atlantic and Mediterranean. Their name refers to the manner in which these lobsters squat against the substrates on which they live. The majority inhabit offshore waters of varying depth and only a few range into the lower intertidal region. Of these Leach's Squat Lobster (*Galathea squamifera*) occurs concealed beneath large stones and in rock pools exposed on the lowest part of shore. The Strigose Squat

Lobster (*Galathea strigosa*) is sometimes seen in public aquaria and is one of the more strikingly coloured species.

Very few species of squat lobsters are commercially fished. Of the native species only the Rugose Squat Lobster (*Munida rugosa*) is occasionally offered for sale in fishmarkets when caught in sufficient numbers.

9.2 COMMERCIALLY IMPORTED SQUAT LOBSTERS

The Yellow Squat Lobster (*Cervimunida johni*) and Red Squat Lobster (*Pleuroncodes monodon*), from Pacific waters of South America are imported occasionally into UK markets; these are discussed in chapter 11.

9.3 LIFE HISTORY OF SQUAT LOBSTERS

The female carries the eggs attached to oosetae of the pleopods. These hatch as shrimp-like larvae [zoeae, fig. 9.1, A] that moult through four or five stages before developing into the benthic megalopal stage (C) that lives on the sea bed. The zoeae of many squat lobster species can be recognised by the combined features of denticulations on the carapace posterior processes and by having a narrow median cleft to the telson (fig. 9.1, B).

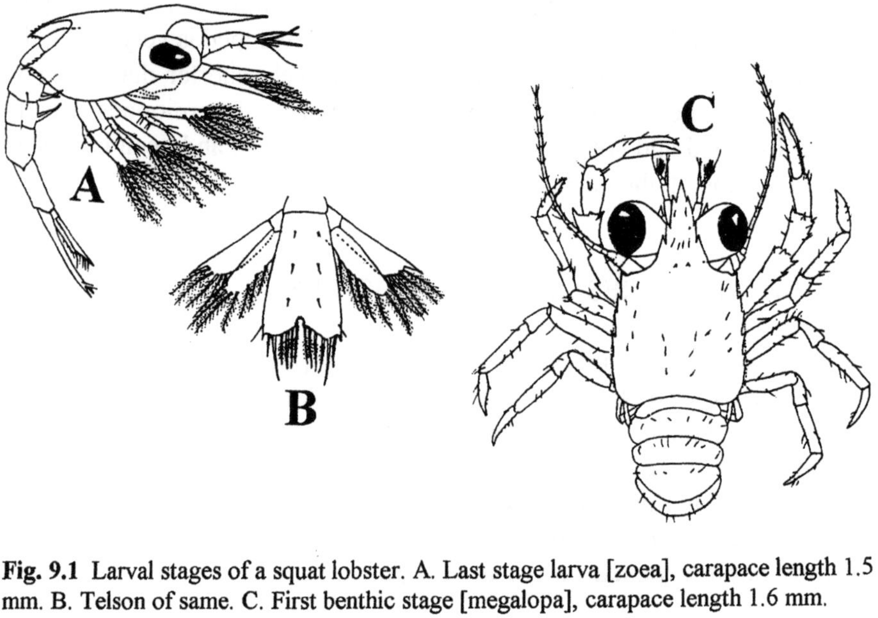

Fig. 9.1 Larval stages of a squat lobster. A. Last stage larva [zoea], carapace length 1.5 mm. B. Telson of same. C. First benthic stage [megalopa], carapace length 1.6 mm.

9.4 BIONOMICS OF SQUAT LOBSTERS

The biology of Squat Lobsters is relatively poorly known. Deep water species (eg. *Munida*) occur on slimy mud to sand substrates and many of those that live in shallow water (*Galathea*) inhabit rocky to algal covered areas or regions containing extensive sea moss (bryozoan) colonies.

9.5 IDENTIFICATION OF SQUAT LOBSTERS

The twenty four species of native squat lobsters are represented by two families. (1). The Chirostylidae: these are all deep water and have the a single transverse suture on the telson. The latter is sharply deflected and carried tucked beneath the last abdominal segment. No members of this family are included in this book. (2). The Galatheidae: species of this family range from shallow to deep water. The telson has numerous sutures and is often lobed. This family contains species of the deep water genus *Munida* (fig. 9.4) that are easily recognised by the spiniform rostrum and the genus *Galathea* (figs 9.2, 9.3) that have a broad flattened rostrum armed with teeth.

Eleven species of *Galathea* inhabit the North Eastern Atlantic/Mediterranean region. Of these the two most frequently seen in captivity are Leach's Squat Lobster (*G. squamifera*), and the Strigose Squat Lobster (*Galathea strigosa*), both are illustrated here. The first mentioned is recognised by the scaly tubercles covering the chelae that are armed with spines only on the margins. The latter species has distinctive red and blue banding on the carapace. Two further species are also not uncommon in some regions: Lilljeborg's Squat Lobster (*Galathea intermedia*) with a broad rostrum and, in the Mediterranean, Zariquiey's Squat Lobster (*G. bolivari*), with a narrow rostrum. Both species have only two spinous processes on the basal peduncular segment of the second antenna compared with the three (as shown fig. 9.2, C for *G. squamifera*) present in the other species mentioned above. For further information about the identification of squat lobsters the reader should consult the relevant references in the 'Further Reading' (p.114).

Galathea squamifera Leach, 1815
Fig. 9.2.
COMMON SYNONYMS: *Galathea glabra.*
DERIVATION: Galatea (Gr) a sea nymph, daughter of Nereus and Doris; squama (squamatus) (Lt)-scaly, alluding to scale covered chelipeds.
VERNACULAR NAMES: Leach's Squat Lobster or Plated Lobster (UK); Troldhummer (Sc); Sastre Negro (Sp).
DISTINCTIVE FEATURES: rostrum (**rs,** fig. A) broad and dorso-ventrally flattened, lateral margins (**lm**, fig. B) armed with three teeth; first antenna-basal segment of peduncle (fig. C) with three spinous processes; chelae (**ce**) covered with scaly tubercles; spines present *only along margins* of chelae propodus (**pro**); fifth pair of pereiopods (**pr5**) small and thin, usually partly or wholly concealed beneath posterior margin of carapace.
COLOUR: carapace and limbs greenish to greenish-brown sometimes spotted with red.
TOTAL LENGTH: 6.0 cm.
DISTRIBUTION: North Eastern Atlantic-from northern Norwegian waters southward to Cape Verde and Azores. The lobster occurs throughout the whole Mediterranean.

Leach's Squat Lobster is widely distributed on North Eastern Atlantic coasts from the lower shore, under stones, and downwards to about 180 m. In the Mediterranean it occurs from depths of approximately 0.5 to 72 m.
GENERAL INFORMATION: this lobster is often common in late spring to early summer on UK coasts where females bearing eggs occur from January until May.

In the Mediterranean this and other species are sometimes found entangled in fishing nets of moored vessels.

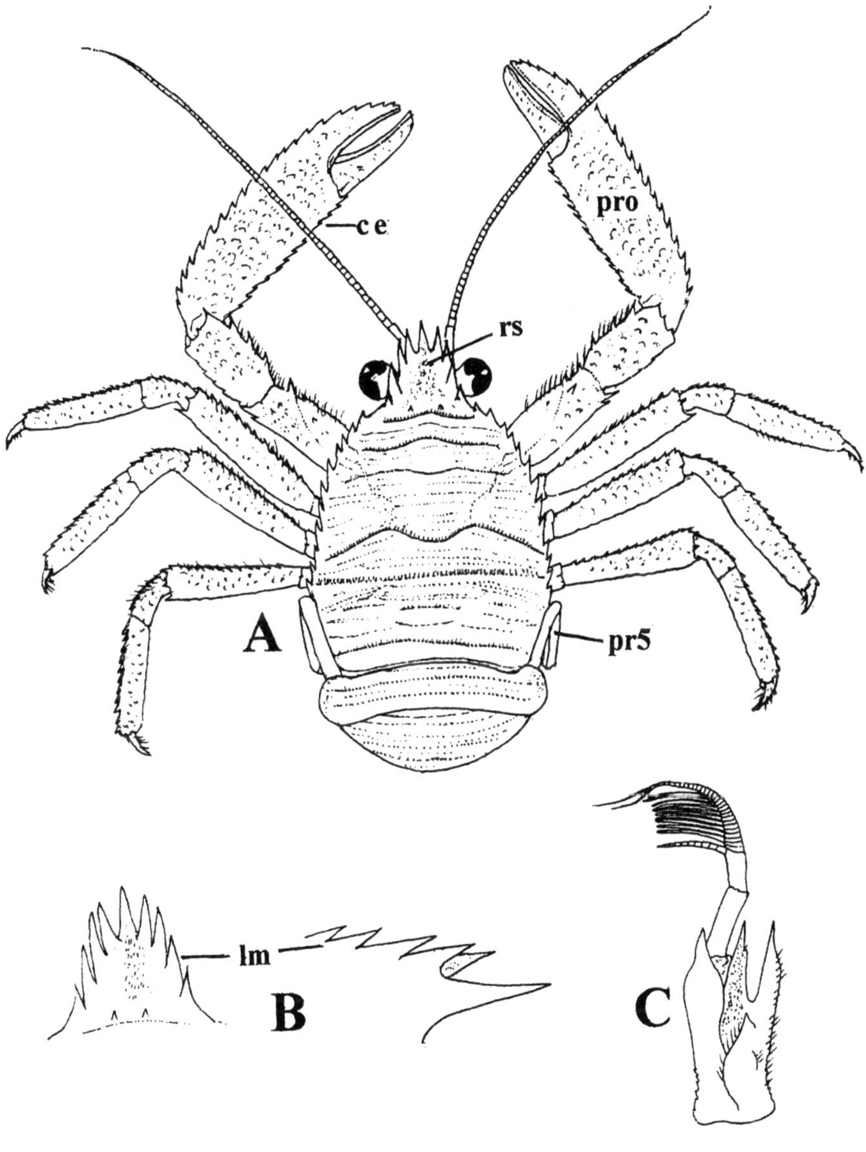

Fig. 9.2 *Galathea squamifera*, Leach's Squat Lobster. A. Whole lobster from dorsal aspect, shown with abdomen folded beneath cephalothorax. B. Rostrum from dorsal (left fig.) and right lateral (right fig.) aspects. C. Left first antenna from ventral aspect. For explanation of annotations see opposite page.

Galathea strigosa (Linnaeus, 1767)
Fig. 9.3.
COMMON SYNONYMS: none.
DERIVATION: Galatea (Gr)-a sea nymph, daughter of Nereus and Doris; strigose (Ni)-beset with striae.
VERNACULAR NAMES: Strigose Squat Lobster, Spiny Squat Lobster (UK).
DISTINCTIVE FEATURES: rostrum (**rs,** figs A, B) broad, dorso-ventrally flattened, apical spine (**sp**, fig. B) long, margins armed with three well defined spines; first antenna-basal segment of peduncle armed with three spinous processes; chelae (**ce**) with conspicuous spines on mid-dorsal surfaces of propodus (**pro**) in addition to spines along margins; fifth pair of pereiopods (**pr5**) very small and thin, usually partly or wholly concealed beneath posterior margin of carapace.
COLOUR: carapace brownish with transverse lines of conspicuously bright blue; extremities of spines and distal parts of dactyl of all limbs and of chelae propodus orange/red.
TOTAL LENGTH: maximum about 14 cm.
DISTRIBUTION: North Eastern Atlantic-from coasts of Scandinavia southwards to Azores and Canary Islands and whole of the Mediterranean.

The Strigose Squat Lobster occurs from the lower shore locally, beneath large stones and boulders downwards to 600 m.

GENERAL REMARKS: this lobster is probably somewhat restricted in distribution on UK coasts and is less common than Leach's Squat Lobster.

Females with eggs occur from January until April.

The blue bandings immediately distinguish Leach's Squat Lobster from all other native species.

The Strigose Squat Lobster is the most conspicuously coloured species of squat lobster in the North Eastern Atlantic and one of the largest species seen in captivity in public aquaria.

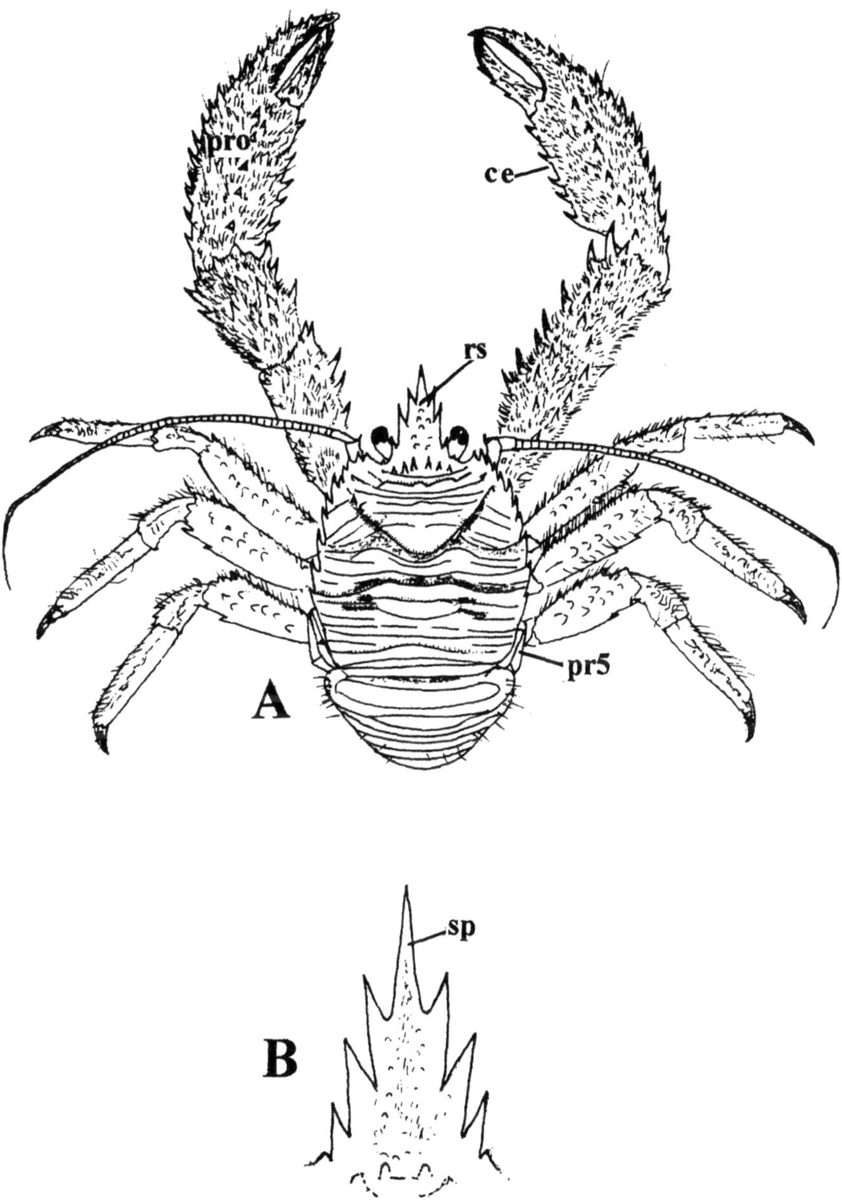

Fig. 9.3 *Galathea strigosa*, Strigose Squat Lobster. A. Whole lobster from dorsal aspect, shown with abdomen folded beneath cephalothorax. B. Rostrum from dorsal aspect. For explanation of annotations see opposite page.

Munida rugosa (Fabricius, 1775)
Fig. 9.4.
COMMON SYNONYMS: *Munida bamffica.*
DERIVATION: probably from munitus (Lt)-armed, alluding to conspicuous spines on rostral region; rugulosus (Lt)-referring to rugosed or wrinkled appearance of carapace.
VERNACULAR NAMES: Rugose Squat Lobster, Plated Lobster, Squat Lobster (UK); Trollhummer (Nw).
DISTINCTIVE FEATURES: flagellum (**fl**) of first antenna very long; chelae (**ce**) of first pair of pereiopods (**pr1**) very long and with numerous spines; rostrum (**rs**) and supra-orbital spines (**spr**) long and thin; carapace and abdomen (**cr**) & (**ab**) with numerous transverse striae that are uninterrupted posteriorly; fourth abdominal segment (**ab4**) without spines; fifth pair of pereiopods (**pr5**) long and thin, usually partly or wholly concealed beneath postero-lateral margins of carapace.
COLOUR: orange to reddish brown; transverse striae on carapace and abdomen darker. Spines and distal parts of chelae white, but spines of chelae deep red with brilliant white tips.
TOTAL LENGTH: maximum about 10 cm.
DISTRIBUTION: North Eastern Atlantic-from Norway southward to Madeira and whole of the Mediterranean.

The Rugose Squat Lobster occurs in depths from about 30-300 m on muddy and sandy/mud substrates in which it forms burrows.

GENERAL INFORMATION: females carrying eggs occur from January until April in British southern waters.

This lobster is sometimes taken in quantity in fish trawls when catches are occasionally marketed.

Munida rugosa is distinguished from the other three species of this genus inhabiting the North Eastern Atlantic (ie. *M. intermedia*, *M. sarsi* and *M. tenuimana*) in having the transverse striae on the posterior part of the carapace uninterrupted and no spines on the fourth abdominal segment.

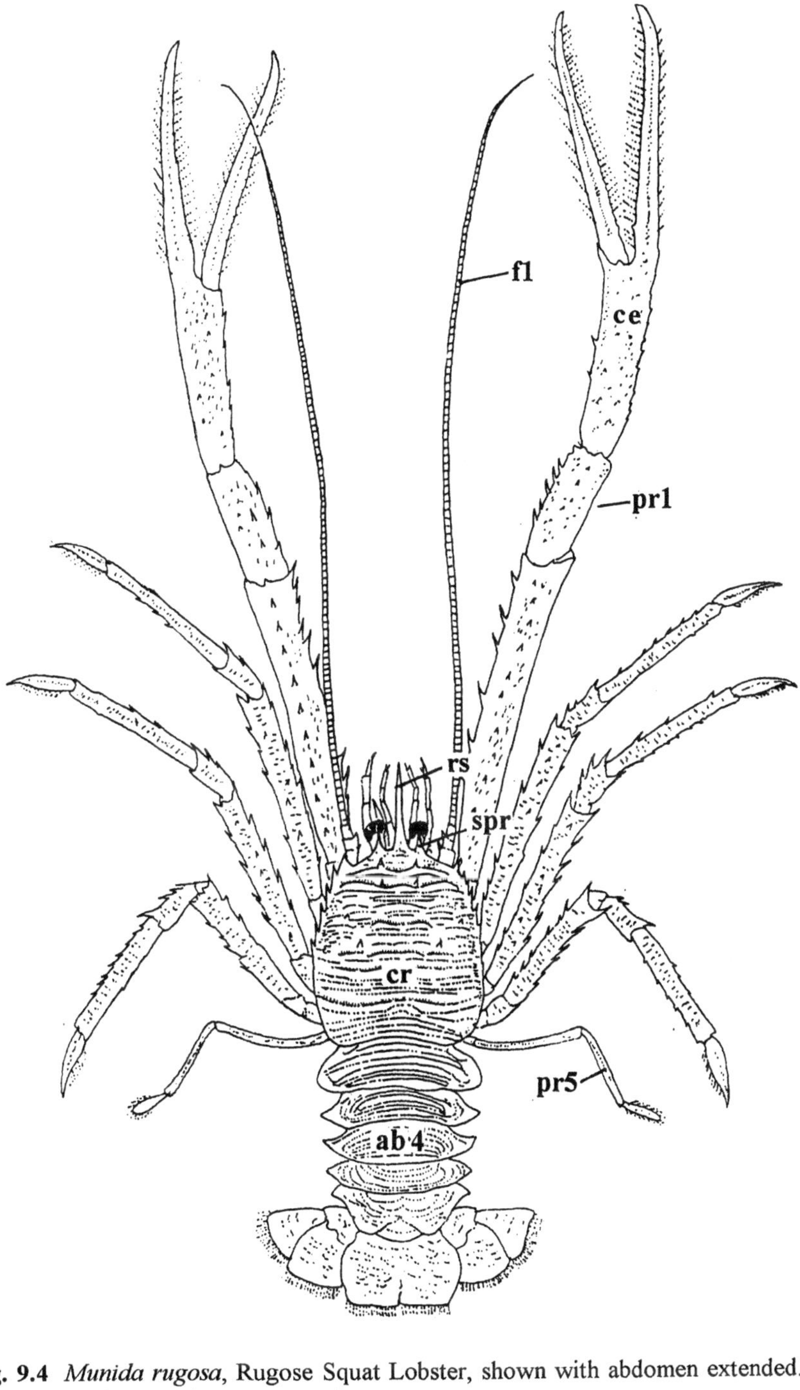

Fig. 9.4 *Munida rugosa*, Rugose Squat Lobster, shown with abdomen extended. For explanation of annotations see opposite page.

9.6 NATIVE SPECIES OF PORCELAIN CRABS

Four species of porcelain crabs inhabit the North Eastern Atlantic and Mediterranean Sea. The Broad Clawed Porcelain Crab (*Porcellana platycheles* and the Long Clawed Porcelain Crab (*Pisidia longicornis*) occur in both areas, whereas *Pisidia bluteli* and *P. longimana* (neither of which have vernacular names) are confined to the Mediterranean region.

9.7 LIFE HISTORY OF PORCELAIN CRABS

Porcelain crabs probably breed throughout most of the year. The eggs are incubated attached to oosetae on the female's pleopods. The larva [zoea, fig. 9.5, A] that hatches is pear-shaped with remarkably long spines that possibly act as buoyancy devices during these planktonic stages. Two further zoeal stages are passed through of increasing size and the last of these moults to the benthic megalopal (B) form. The juvenile crab stage that follows adopts a life-style similar to the adult.

Fig.9.5 Larval stages of a porcelain crab. A. First stage larva [zoea], total length 7.5 mm. B. First benthic stage [megalopa], carapace length 2.5 mm.

9.8 BIONOMICS OF PORCELAIN CRABS

Porcelain crabs are commonly found on the middle to lower shore beneath stones, rocks and weed cover. They can cling to solid surfaces with considerable tenacity using the sharp curved dactyls of the pereiopods.

When seized they will shed their limbs more readily than most other crabs and care must be taken when detaching these crabs from rock surfaces.

All Porcelain crab species are far too small to be considered as edible crustaceans.

9.9 IDENTIFICATION OF PORCELAIN CRABS

Obvious feature that allow the native species to be distinguished one from the other are as follows. The Broad Clawed Porcelain Crab (*Porcellana platycheles*) is recognised from all others by having flat, broad and hairy chelae (**ce**, fig. 9.6) and the carapace margins unarmed. Of the *Pisidia* species the penultimate segment of the first antenna peduncle of *P. bluteli* has a spine and the merus to propodus of the second pair of pereiopods has spines on the anterior margins. *P. longimana* has spines on the dorsal surface of the carapace behind the cervical groove; these are absent in *P. longicornis* (fig. 9.7).

Porcellana platycheles (Pennant, 1777)
Fig. 9.6.
COMMON SYNONYMS: none.
DERIVATION: porcellan (It)-the Venus shell or nacre (mother or pearl) of the shell-alluding to appearance of crab's carapace; platys (Gr)-flat; chele (Gr)-a claw-referring to the flat chelipeds.
VERNACULAR NAMES: Broad Clawed Porcelain Crab (UK); Porselen-skrabber (Nw); Porcelaenskrabbe (Dm); Harig Porceleinkrabbetje (Nd).
DISTINCTIVE FEATURES: carapace (**cr**) flat and covered with long hairs extending onto large and flattened chelae (**ce**) and following limbs giving an overall 'shaggy' appearance; lateral margins (**lm**) of carapace without spines (antero-and postero-carapace lateral margins not distinct one from the other); fifth pair of pereiopods (**pr5**) very small and usually concealed beneath posterior margin of carapace.
COLOUR: carapace a mother-of pearl colour and texture, but surface obscured by brownish to greenish brown hairs often matted with mud from substrate.
CARAPACE LENGTH: maximum about 1.6 cm.
DISTRIBUTION: North Eastern Atlantic-from the Shetland Islands southwards to the Canary Islands and whole of the Mediterranean.

The Broad Clawed Porcelain Crab is widely distributed from the middle to lower shore and downwards to about 6 m depth.

GENERAL INFORMATION: this crab is often found clinging to undersides of rocks or on muddy sand substrates.

Females bearing eggs occur from March until August on south coasts of the British Isles and from July until September in parts of the Mediterranean.

Substrate particles adhering to the setal mat covering body and limbs provide an effective camouflage and crabs can be easily overlooked on muddy substrates.

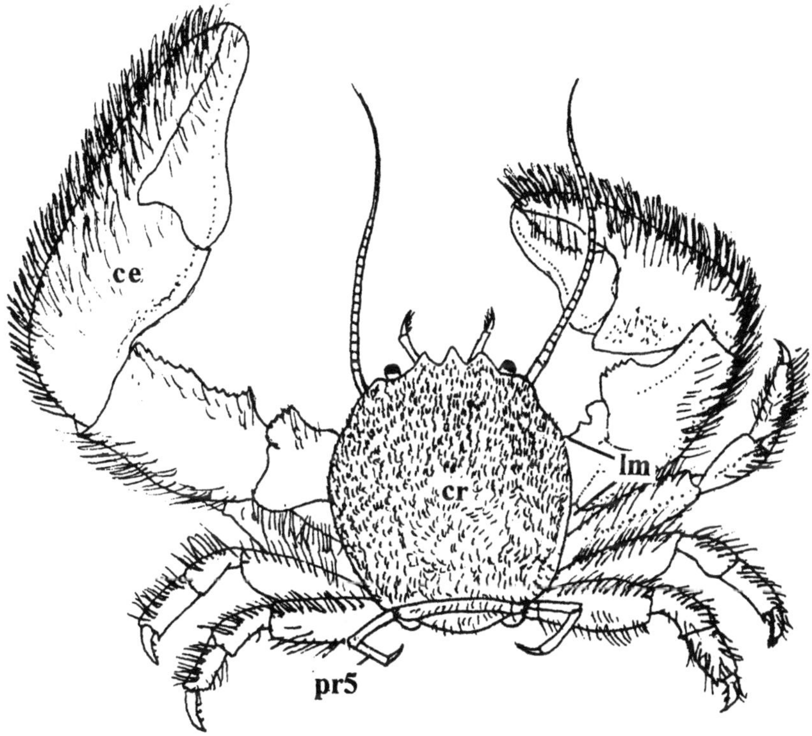

Fig. 9.6 *Porcellana platycheles*, Broad Claw Porcelain Crab. For explanation of annotations see opposite page.

Pisidia longicornis (Linnaeus, 1767)
Fig. 9.7.
COMMON SYNONYMS: *Porcellana longicornis.*
DERIVATION: the naturalist William Elford Leach who established the genus *Pisidia* may have derived the name from Pisidia-an inland district of Asia Minor north of Lycia and Pamphylia, alternatively he may have construed it from pisos (Gr)-a pea alluding to the rounded body. The name longicornis however, as give by Linnaeus, i.e. longus (Lt)-long, and corn (Lt)-a horn is difficult to explain. Perhaps he was alluding to the calcified long chelipeds of this species.
VERNACULAR NAMES: Long Clawed Porcelain Crab (UK); Porceleinkrabbetje, Zespoot, Langspriet (Nd).
DISTINCTIVE FEATURES: carapace (**cr**) suboval dorsally and sparsely setosed at most; lateral margins (**lm**) of carapace with indistinctly developed teeth (antero-and postero-lateral margins not distinct one from the other); chelae (**ce**) long; a few club-shaped setae on second to fourth pereiopods (**pr2-pr4**), fifth pair of pereiopods (**pr5**) very small and usually concealed beneath posterior margin of carapace.
COLOUR: reddish/orange with spots of black, sometimes part or whole of carapace dorsal surface pearly white.
CARAPACE LENGTH: maximum about 1 cm.
DISTRIBUTION: North Eastern Atlantic-from Norway southwards to Angola, West Africa and whole of the Mediterranean.

The Long Clawed Porcelain Crab occurs from the lower shore downwards to a depth of about 70 m in North Eastern Atlantic but more offshore (e.g. 30 to 100 m) in some Mediterranean localities.
GENERAL INFORMATION: females bearing eggs occur from March until August on southern coasts of the British Isles and from February until September in the Mediterranean.

In some European fishmarkets these crabs are sometimes found mixed with mussels and oysters.

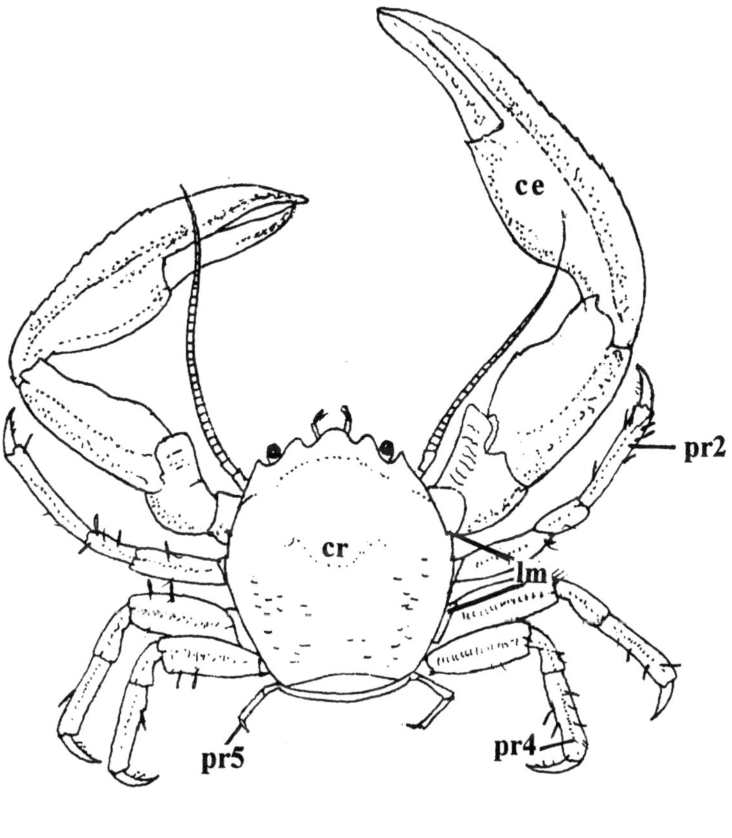

Fig. 9.7 *Pisidia longicornis*, Long Clawed Porcelain Crab. For explanation of annotations see opposite page.

FURTHER READING

Allen, J.A. (1967) *The Fauna of the Clyde Sea Area. Crustacea: Euphausiacea and Decapoda.* Scottish Marine Biological Association, Millport (pages 57-59 and 91, 94, 95 for squat lobsters and porcelain crabs).

Crothers, J. and Crothers, M. (1983) A key to the crabs and crab-like animals of British inshore waters. *Field Studies*, **5**, 753-806 (pages 760-761 and 787 for porcelain crabs).

Rice, A.L. and de Saint Laurent, M. (1966) The nomenclature and diagnostic characters of four north-eastern Atlantic species of the genus *Munida* Leach: *M. rugosa* (Fabricius), *M. tenuimana* G.O. Sars, and *M. sarsi* Huus, (Crustacea, Decapoda,Galatheidae). *Journal of Natural History*, **20**, 143-163.

Zariquiey Alvarez, R. (1968) Crustáceòs Decápodos Ibéricos. *Investigación Pesquera,* **32**, 1-510 (pages 261-293 for squat lobsters and porcelain crabs).

10

True Crabs (Brachyura)

10.1 NATIVE SPECIES

At least one hundred and forty species of true crabs inhabit the North Eastern Atlantic Ocean and Mediterranean Sea. The twenty seven species described and figured are those most likely to be seen on seashores, in sea life centres, among quayside landings of fishes etc., and for sale in fishmarkets.

10.2 INTRODUCED AND IMMIGRANT SPECIES

At least five species of non-indigenous brachyuran crabs have been reported from various localities in the British Isles. The South American xanthid crab (*Pilumnoides inglei*) was discovered attached to hulls of ships during the early part of this century but has not been found since. The East American Caribbean Xanthid Crab (*Dyspanopeus sayi*) and the Rock Crab (*Brachynotus sexdentatus*) from the Mediterranean region, became established in a warm water dock in Wales during the 1950-1960's, but only the Caribbean Xanthid Crab survived and continued breeding after a fall in dock temperature when the adjacent power station was closed. There are now several records of the American Chesapeake Bay Swimming Crab (*Callinectes sapidus*) from coastal waters of southern, eastern England and the Netherlands and there are well established breeding populations of this species in the eastern Mediterranean. The Dwarf Xanthid Crab (*Rhithropanopeus harrisii*) of the American Atlantic occurs in the Netherlands, and the far eastern Mitten Crab (*Eriocheir sinensis*) has been breeding in Europe for well over half a century and is probably now established in some British localities.

At least eleven Indo-Pacific species of true crabs have migrated into the

Mediterranean via the Suez Canal. These are represented by two nut crabs (*Leucosia signata* and *Myra fugax*), six species of swimming crabs (Blue Swimming Crab-*Portunus pelagicus*, *Thalamita poissonii*, *Charybdis longicollis*, *C. helleri*, *C. sexdentatus* and *C. merguiensis*), two xanthid crabs (*Heteropanope laevis* and *Pilumnopeus vauquelini*) and one goneplacid Crab (*Eucrate crenata*). Except for the Blue Swimming Crab, none of the others have common names.

10.3 NATIVE SPECIES OF ECONOMIC IMPORTANCE

10.3.1 The European Edible Crab

The European Edible Crab (*Cancer pagurus*) provides an important source of income for local fishermen. The major fishing regions are off the coasts of Norway, Scotland, north east and south west England and also Brittany. Smaller local fisheries are to be found in many other countries whose coasts border the north eastern Atlantic.

(a) Fishing methods

During the early part of the nineteenth century, when European Edible Crabs were plentiful in the intertidal regions, they were often fished using a gaff to pull them from crevices or by trapping in hoop nets. Nets were often constructed of a mesh bag attached to an iron ring of approximately 60 cm in diameter. The hoop nets were baited with fish, lowered into pools of suitable depths and hauled at intervals to remove crabs. As inshore populations of crabs declined they were then fished in deeper waters with creels and pots of the same design and construction and using the same methods described for fishing clawed lobsters (see 5.4). Hoop net fishing is now rarely seen.

Boats used for local fishing of these crabs vary in design but are usually of shallow draft and suitable for launching from sloping beaches. On the Yorkshire and Northumberland coasts for example, 'keel boats' are used. These are 12-15 m in length, decked, and with an aft wheel-house and an internal diesel engine. They may carry a crew of five who are able to set some 400-500 pots during one fishing excursion.

Fishing is usually governed by weather conditions. In the UK some crab fishing continues throughout the year but many crab fishermen set pots from the months of April through to November. Quantities of crabs caught can vary weekly or monthly. When catches of marketable crabs decrease, (particularly in late summer during the period of moulting), fishermen often change their fishing grounds to inshore waters and set pots closer together to catch lobsters that provide greater financial returns than Edible Crabs.

(b) Marketing European Edible Crabs

Catches are sold alive either by auction at local markets or direct to restaurants, hotels or shellfish processors. Some fishermen retain the whole or part of their catches that are cooked and sold, along with other shellfish, on local stalls or in the shops selling sea produce commonly seen at coastal holiday resorts. Regulations governing the minimum landing sizes and condition of European Edible Crabs vary (see p. 136).

Crabs are usually killed within about 30 minutes when they are immersed in freshwater at a temperature of 38-49°C [100-120°F]. An alternative method is 'spiking,' ie., insertion of an awl slightly above the mouthparts and angled towards the head to destroy the brain. This method is sometimes used to kill large crabs. Freshly killed crabs are placed in a wire mesh basket and immersed in boiling salted water for about twenty to thirty minutes according to the volume of crabs being cooked; they are then removed, hosed with cold water and left to cool. Crabs are graded for size and quality. Unblemished and undamaged crabs are sold whole, the remainder are 'picked' for removal of white meat (the bulk of which comes from the chelae), and the brown meat, (that forms the digestive glands and reproductive organs located beneath the carapace). Extracted white meat is packed whole whereas brown meat is usually minced and added to supplement the white meat. Both forms when packed are frozen preparatory to marketing. Some crab meat is used to produce the various forms of potted paste.

10.3.2 Other comestible species of true crabs

In parts of the Mediterranean there are fisheries for the introduced Chesapeake Bay Swimming Crab (*Callinectes sapidus*) well established in that region. A limited UK export market now exists for the Thorn-back Spider Crab (*Maja squinado*). For many years this crab was usually discarded by British fishermen but was always eaten in moderate quantities by many Europeans. The Velvet Swimming Crab (*Necora puber*) and North Atlantic Shore Crab (*Carcinus maenas*) is sold in some French markets, and the Mediterranean Shore Crab (*Carcinus aestuari*) and the Warty Xanthid Crab (*Eriphia verrucosa*) is fished and eaten by Mediterranean communities. Claws of the Moroccan Fiddler Crab (*Uca tangeri*) are consumed in Spain and Portugal and occasionally marketed in Europe. The Sardine Swimming Crab (*Polybius henslowii*) is still used in Portugal as a land fertilizer when sufficient quantities are caught by sardine fishermen.

Very few exotic crab species are imported into UK or European markets. Large specimens of the Blue Swimming Crab (*Portunus pelagicus*), the Mud Swimming Crab (*Scylla serrata*), both from the Indopacific, and the Atlantic Snow Spider Crab (*Chionoecetes opilio opilio*) are occasionally offered for sale (see chapter 11). The introduced Mitten Crab (*Eriocheir sinensis*) is also eaten by European oriental communities.

Soft shelled crabs (termed peelers), especially the North Atlantic Shore Crab (*Carcinus maenas*) are often used for bait by anglers and newly moulted Mediterranean Shore Crabs (*Carcinus aestuari*) are sometimes seen in markets of that region, particularly in Venice.

10.4 LIFE HISTORY

Females of many true crab species are receptive to mating after they have moulted and when the shell is quite soft. At this time sex-attractant chemicals [pheromones] are released. The male of the North Atlantic Shore Crab, for example, carries the female clasped beneath his body shortly before she is due to moult. The female may be totally or partly released to allow moulting and is then repositioned ventral side uppermost for copulation to take place. The European Edible Crab behaves similarly and the male may remain with his partner until her shell has partially hardened. Female North Atlantic Shore Crabs spawn from 6-17 weeks after mating, this interval may be a few months for the European Edible Crab.

The eggs of true crabs hatch as 'comma'-shaped planktonic larvae [zoeae] (fig. 10.1, A) except those of sponge crabs in which the zoeae are shrimp-shaped. Further stages of similar shapes follow, one for spider crabs to as many as seven for the Chesapeake Bay Swimming Crab, until the benthic stage [megalopa] (fig. 10.1, B) is reached. This lives on the sea bed where it moults to the juvenile crab stage.

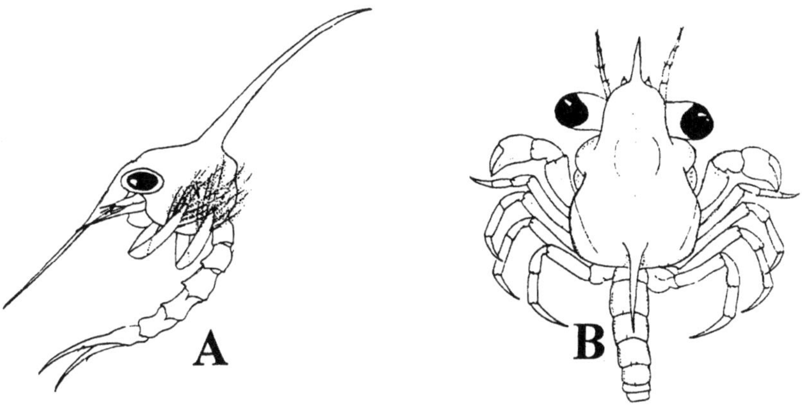

Fig. 10.1 Larval stages of a true crab. A. First zoeal stage, total length 1.38 mm. B. megalopal stage, carapace length 1.3 mm.

10.5 LIFE STYLES

True crabs show considerable diversity of life styles. The paddle-shaped fifth pair of pereiopods evolved by swimming crabs are used to attain considerable swimming speeds and the somewhat flattened body presents minimum resistance when these crabs need to rapidly bury themselves into soft substrates when being chased by predators. By comparison, the very flattened body of rock crabs allows them to slip easily into very narrow rock fissures barred to pursuing predators. Slow moving spider crabs conceal themselves with a covering of seaweeds that provides excellent camouflage on weedy substrates. Pea crabs are adapted to living within live mussels or sea squirts.

Intertidal species such as the North Atlantic Shore Crab and small specimens of the European Edible Crab are some times abundant among stones on mid-to low shores of many Atlantic coasts. The typical habitat for medium to large specimens is beneath weed covered rocks. European Edible

Crabs of small to medium size often occur buried in sand at the edge of large boulders.

The Velvet Swimming Crab is common in some regions and often inhabits weedy pools of the lower shore. It is noticeably aggressive and typically faces a threat with claws outstretched revealing bright red markings on the inner surfaces. This posture is also adopted by other crab species and may momentarily confuse predators by giving an illusion of increased body size and change of shape.

On southern Atlantic beaches of Spain, Portugal and Mediterranean countries Moroccan Fiddler Crabs are commonly seen scurrying over intertidal muddy sand surfaces that are marked with their burrows.

10.6 IDENTIFICATION

The species of true crabs described in this book belong to fifteen families. The more obvious features of each family (as they obtain for species listed in the systematic classification on pp17-19) are briefly described below.

1. Sponge crabs (Family Dromiidae): the body is sub-globular and the posterior two pairs of pereiopods reflexed for carrying a sponge over the carapace.

2. Box crabs (Family Homolidae): recognised by the distinctive long eyestalks, a box-shaped spinose carapace and long thin pereiopods.

3. Demon-faced crabs (Family Dorippidae): the two posterior pairs of pereiopods are strongly reflexed, dorsally placed, and used for carrying a discarded shell of a bivalve mollusc; the carapace dorsal surface has a 'human face-like' configuration.

4. Nut crabs (Family Leucosiidae): the carapace is globular (nut-shaped) and the eyes are very small and well concealed.

5. Helmet crabs (Family Corystidae): the carapace is a distinctive helmet-shaped, the chelae are long and thin and the antennal flagella long; these crabs live most of the time just beneath sand surfaces.

6. Circular crabs (Family Atelecyclidae): the carapace is almost circular with strongly toothed margins.

7. Edible crabs (Family Cancridae): the carapace is an oval shape with a 'pie-crust edge' pattern on the antero-lateral margins; these crabs are fished for human consumption.

8. Geryonid crabs (Family Geryonidae): the carapace is a hexagonal-shape

bearing antero-lateral teeth and the second to fifth pairs of pereiopods are relatively long and thin.

9. Swimming crabs and shore Crabs (Family Portunidae): these are probably the commonest members of all true crabs. All have a sub-oval carapace with well developed teeth on the antero-lateral margins. Swimming crabs have a distinctive flattened fifth pair of pereiopods, the terminal segment (dactylus) of which is 'paddle-shaped' in most species.

10. Xanthid crabs (Family Xanthidae): there are no obvious features common to the whole family. The species described have a hexagonal shaped carapace that is very spinosed (Warty Xanthid Crab-*Eriphia*), or with rounded marginal lobes (*Xantho* species) or with a covering of patches of hair (*Pilumnus* species).

11. Pea crabs (Family Pinnotheridae): usually have a soft oval-shaped carapace and live within bivalve molluscs (commonly mussels) and ascidians (sea squirts).

12. Rock crabs and mitten crabs (Family Grapsidae): the carapace is somewhat squarish and noticeably flattened. The shore dwelling species move very quickly over rocks, harbour walls or pier support surfaces on which they are often seen. Mitten crabs inhabit rivers.

13. Fiddler crabs (Family Ocypodidae): the eyestalks are long. Males have one cheliped very enlarged. The entrances of their burrows on sand and mud is usually marked by small mounds of the substrate.

14. Freshwater crabs (Family Potamidae): the carapace branchial regions are slightly inflated. These crabs inhabit freshwater streams.

15. Spider crabs (Family Majidae): the carapace is triangular. Many species have long thin pereiopods and the body is often covered with pieces of seaweed or other encrustations. They are slow moving crabs.

Dromia personata (Linnaeus, 1758)
Fig. 10.2.
COMMON SYNONYMS: *Dromia vulgaris.*
DERIVATION: dromaios (Gr)-swift; personatus (Lt)-masked-alluding to masking of the carapace with a sponge. The naturalist Freidrich Weber established the generic name *Dromia* for this crab but why he chose this name for a particularly lethargic species is difficult to comprehend.
VERNACULAR NAMES: Linnaeus's Sponge Crab, Sleepy Crab, Little Hairy Crab (UK); Wolkrab (Nd); Dromie, Crabe Dormeur (Fr).
DISTINCTIVE FEATURES: carapace broadly octagonal in outline; dorsal surfaces of carapace and limbs covered with fine short brown pubescence; antero-lateral region (**al**) cut into five teeth on each side; inner surfaces of dactylus (**dcy**) and propodus (**pro**) of chelae (**ce**) with long dense mat of hairs; fourth (**pr4**) and fifth (**pr5**) pairs of pereiopods dorsally placed and with small claws (**cl**).
COLOUR: dense pubescence on body dark brown to golden brown, naked tips of chelipeds pale pink or white.
CARAPACE LENGTH: maximum 7.5 cm.
DISTRIBUTION: North Eastern Atlantic-from southern North Sea southwards to Canary Islands. Probably whole of Mediterranean.

Linnaeus's Sponge Crab rarely occurs in pools on the lowest part of the shore and is distributed downwards to 100 m depth. Often occurs on rocky substrates.

GENERAL INFORMATION: females with eggs occur in July and August.

The Sponge Crab cuts pieces of sponge or removes a whole sea squirt from a main colony and shapes these items to fit over the carapace. This camouflage 'hat' is held in position by the specially modified claws of the fourth and fifth pereiopods.

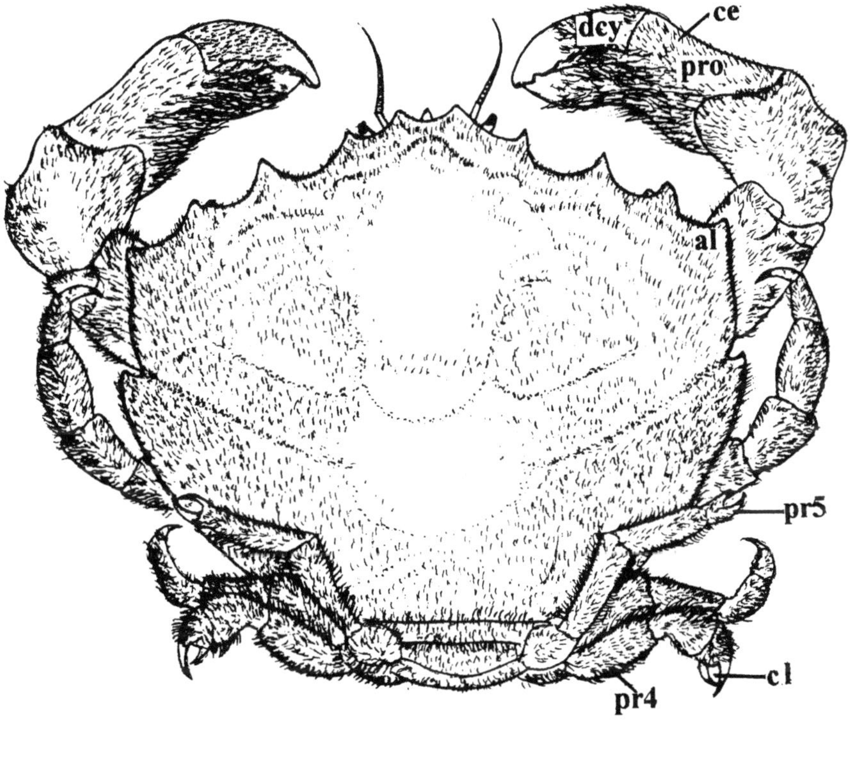

Fig. 10.2 *Dromia personata*, Linnaeus's Sponge Crab. For explanation of annotations see opposite page.

Paromola cuvieri (Risso, 1816)
Fig. 10.3.
COMMON SYNONYMS: none.
DERIVATION: para (Gr)-near or showing affinities with the genus *Homola* (see following species).
VERNACULAR NAMES: Box Crab (UK); Hãxkrabbe (Sw); Paromole (Fr).
DISTINCTIVE FEATURES: carapace (**cr**) approximately rectangular, surface invested with conspicuous spines of various sizes, rostrum (**rs**) prominent; antero-(**al**)-and postero-(**ps**) lateral margins strongly curved, sides deep; eyes (**ey**) exposed and with two-segmented eyestalks, distal segment swollen; pereiopod surfaces with spines, first to fourth (**pr1-pr4**) long and thin, very spinosed, fifth pair (**pr5**) terminating in a partly formed claw (**cl**).
COLOUR: dorsal surface of carapace and pereiopods reddish brown, distal parts of dactylus and cheliped propodus dark brown to black.
CARAPACE LENGTH: maximum about 21.5 cm.
DISTRIBUTION: North Eastern Atlantic-from Iceland southwards to Angola (W Africa). Whole of Mediterranean.

The Box Crab is predominantly a deep-water species ranging from about 150 m to over 1000 m in depth and on muddy to coral substrates. It occasionally occurs in shallow coastal waters.

GENERAL INFORMATION: females with eggs occur from March and probably throughout the year until November.

This crab is sometimes captured by vessels fishing offshore waters and landed for curiosity rather than edible qualities.

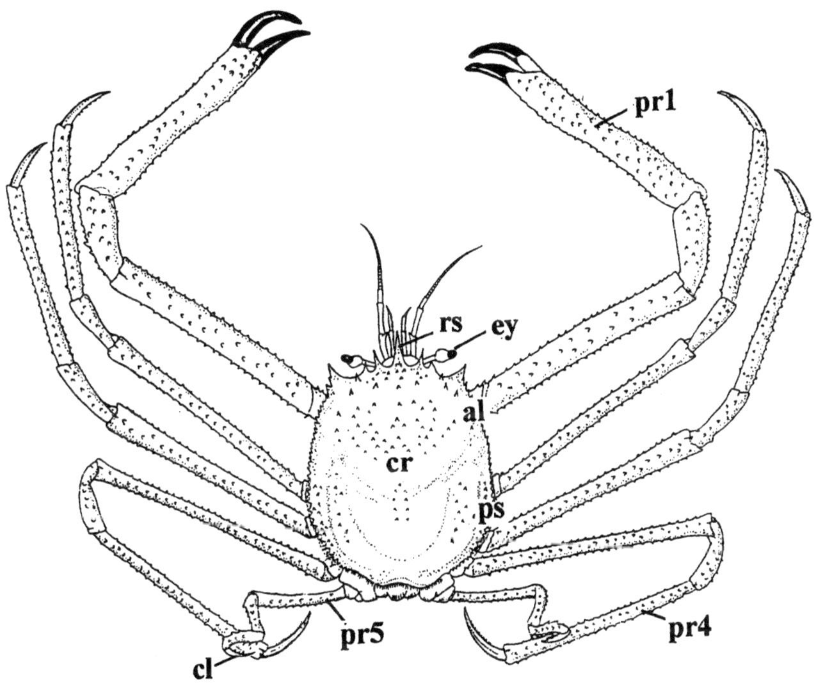

Fig. 10.3 *Paromola cuvieri*, Box Crab. For explanation of annotations see opposite page.

Homola barbata (Fabricius, 1793)
Fig. 10.4.
COMMON SYNONYMS: *Thelxiope barbata.*
DERIVATION: the generic name of this crab was established by the English naturalist William Elford Leach but, as is the case for many crustacean names he proposed, its derivation is not obvious. A suggestion is homolas (Gr)-smooth, even-perhaps alluding to the relatively smooth but granular carapace; barbatus (Lt)-bearded, refers to the hirsute legs.
VERNACULAR NAMES: Hairy Box Crab (UK); Homole (Fr).
DISTINCTIVE FEATURES: carapace rectangular, sides deep, dorsal surface granular and with well developed spines on anterior half, rostrum (**rs**) small and bifurcated at tip; antero-(**al**) and postero-(**ps**) lateral margins straight; second antenna (**an2**) very long; eyes (**ey**) long and exposed, eyestalks two-segmented, distal segment large; chelae (**ce**) robust, surfaces very hairy and granular, second to fourth pereiopods (**pr2-pr4**) long, somewhat flattened and moderately hairy, fifth pereiopod (**pr5**) terminating in a partly formed claw (**cl**).
COLOUR: body and limbs yellowish to reddish brown, spines deeper red.
CARAPACE LENGTH: maximum 3.65 cm.
DISTRIBUTION: North Eastern Atlantic-from Spain to Cape Verde Islands. (South African records may represent a different species). Whole of Mediterranean. This species also occurs on the American Atlantic coast from South of Massachusetts to Brazil.

The Hairy Box Crab occurs in offshore waters usually between approximately 40 and 100 m depth on various substrates.
GENERAL INFORMATION: females with eggs occur from April and probably through until September.

This crab is often caught in large numbers in traps set for squid etc., and not uncommonly seen on Mediterranean quaysides entangled in fishing nets.

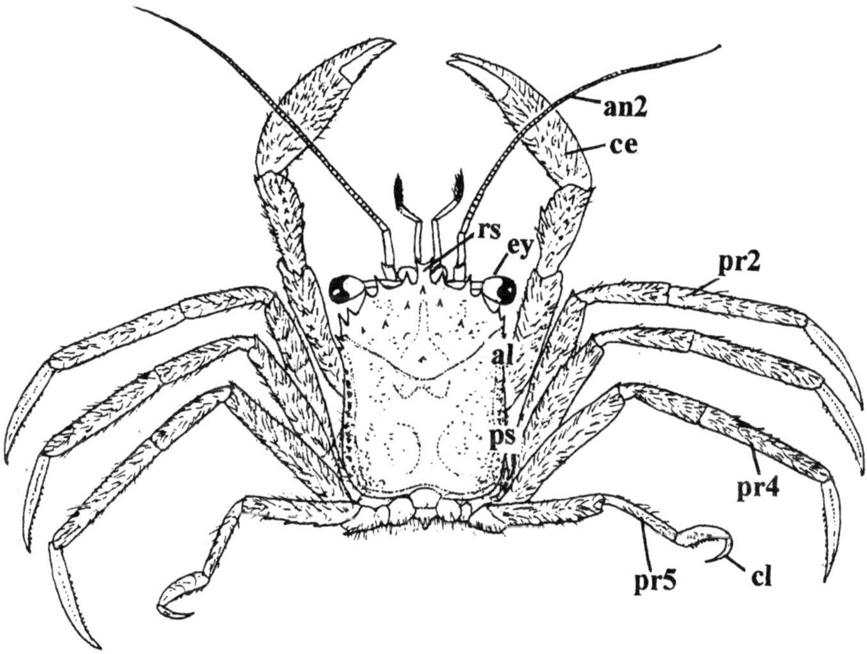

Fig. 10.4 *Homola barbata*, Hairy Box Crab. For explanation of annotations see opposite page.

Medorippe lanata (Linnaeus, 1767)
Fig. 10.5.
COMMON SYNONYMS: *Dorippe lanata.*
DERIVATION: me (Gr)-not (ie., not belonging to genus *Dorippe*); doripon (Gr)-warlike, perhaps alluding to the aggressive nature of this crab, or crab with a 'demon face' sculpturing on carapace; lanata (Lt)-woolly, alluding to hairy margins of carapace and limbs.
VERNACULAR NAMES: Demon-faced Porter Crab, Porter Crab, Face Crab (UK); Franquet Fuster (Sp).
DISTINCTIVE FEATURES: dorsal surface of carapace (**cr**) sculptured with configuration of a human face, antero-lateral margins (**al**) conspicuously diverging posteriorly and separated from postero-lateral margins (**ps**) by a stout spine; merus (**mrs**) of second and third pereiopods (**pr2, pr3**) with spines on anterior margin and with dactylus (**dcy**) long and curved; fourth (**pr4**) and fifth (**pr5**) pereiopods terminating in small partly formed claw (**cl**); male with propodus (**pro**) of first pereiopod (**pr1**) swollen.
COLOUR: brownish to reddish brown.
CARAPACE LENGTH: maximum about 2.2 cm.
DISTRIBUTION: North Eastern Atlantic-from Northern Portugal southward to Angola (West Africa). Whole of Mediterranean.

The Demon-faced Porter Crab commonly occurs in offshore waters between about 40-50 m depth, and occasionally in excess of 100 m. It inhabits muddy sandy and shelly substrates.

GENERAL INFORMATION: the crab frequently carries a discarded shell of a bivalve mollusc over the carapace, presumably for camouflage. This is held in place by claws of fourth and fifth pereiopods. Hence the vernacular name 'Porter Crab' ie. one who carries a burden.

The crab is sometimes caught in considerable numbers in trawls of fishing vessels and often occurs in squid or crab/lobster pots.

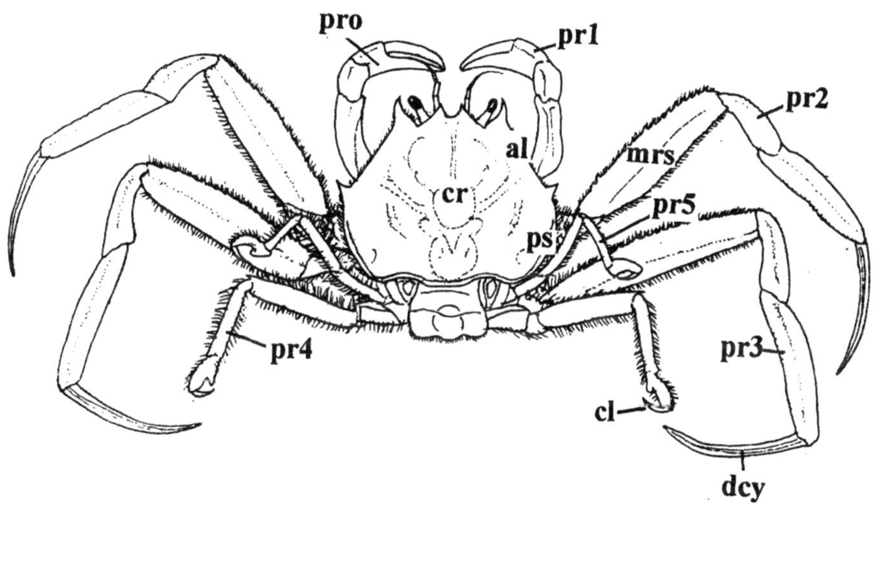

Fig. 10.5 *Medorippe lanata*, Demon-faced Porter Crab, female. For explanation of annotations see opposite page.

Ebalia tuberosa (Pennant, 1777)
Fig. 10.6.
COMMON SYNONYMS: none.
DERIVATION: perhaps derived from ebaios (Gr)-small; tuberosus (Lt)-full of humps.
VERNACULAR NAMES: Pennant's Nut Crab, Pennant's Ebalia, Uneven Crab (UK); Ruwe Kiezelkrab (Nd).
DISTINCTIVE FEATURES: carapace (**cr**) octagonal in outline, dorsal surface with numerous tubercles and with a longitudinal and transverse elevation forming an indistinct cross; antero-lateral margins (**al**) granular; eyes very small; chelipeds (**pr1**) robust, propodus (**pro**) swollen; second to fifth pereiopods (**pr2**, **pr5**) short and stout.
COLOUR: variable, carapace and pereiopods light yellowish orange to deep reddish brown, sometimes spotted with white.
CARAPACE LENGTH: maximum 1.8 cm.
DISTRIBUTION: North Eastern Atlantic-from Norway southward to Canary Islands. Whole of Mediterranean.

Pennant's Nut Crab rarely occurs intertidally and is distributed downwards to over 200 m depth. In the Mediterranean it inhabits waters of about 30-130 m depth.

The crabs occurs on gravel, sandy and muddy substrates, sometimes on shelly sea beds.
GENERAL INFORMATION: females carrying eggs may occur throughout the year in North Eastern Atlantic waters.

The crab normally burrows into the sea bed and rests with only the anterior part of the carapace protruding above the substrate surface.

Many of the *Ebalia* species occurring in the North Eastern Atlantic-Mediterranean waters are not particularly easy to identify and opinions differ as to the taxonomic validity of some of these. At least nine species may be present in these regions. Pennant's Nut Crab is the commonest in many localities and is occasionally seen in public aquaria.

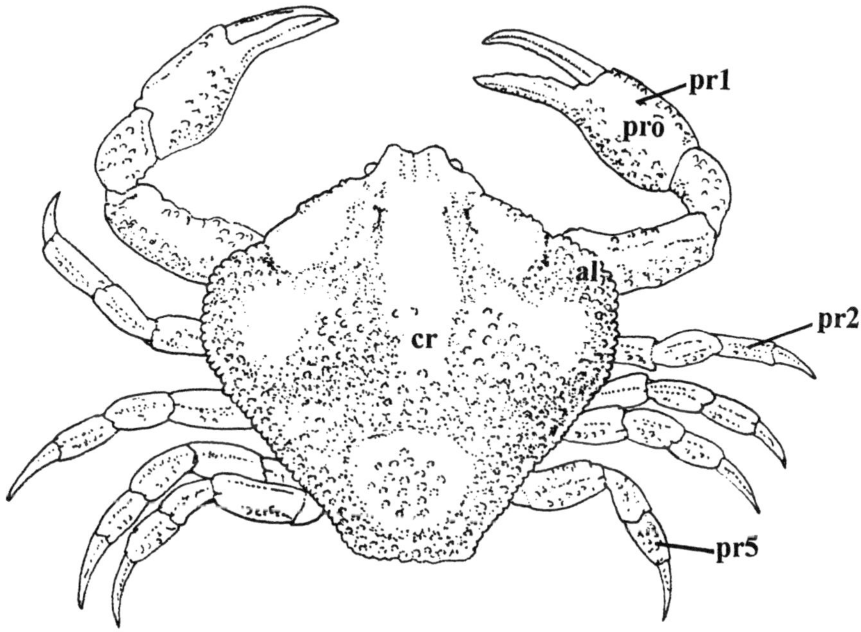

Fig.10.6 *Ebalia tuberosa*, Pennant's Nut Crab. For explanation of annotations see opposite page.

Corystes cassivelaunus (Pennant, 1777)
Fig. 10.7.
COMMON SYNONYMS: none.
DERIVATION: korystes (Gr)-a helmeted man; Cassivellaunus-a British chief defeated by Caesar in 54 BC.
VERNACULAR NAMES: Helmet Crab, Long-clawed Crab, Old Man's Face Crab, Masked Crab (UK); Maskekrabbe (Sc); Helmkrab (Nd); Crabe Masqué (Fr); Cangrejo Dentado (Sp).
DISTINCTIVE FEATURES: carapace (**cr**) longitudinally oval in shape with three teeth on each antero-lateral (**al**) margin and one on hind part of postero-lateral (**ps**) margin; second antenna (**an2**) long (especially in male); chelipeds (**pr1**) long and stout; second (**pr2**) to fifth (**pr5**) pereiopods somewhat flattened; male's chelipeds longer than those of female's.
COLOUR: carapace dorsal surface pale mauve to reddish brown; pereiopods yellowish to white.
CARAPACE LENGTH: maximum about 4 cm.
DISTRIBUTION: North Eastern Atlantic-from Scandinavia southward to Gibraltar. Whole of Mediterranean.

In the North Eastern Atlantic the Helmet Crab occasionally occurs just below the sand surface exposed at low water spring tides and is found downwards to about 115 m depth. In the Mediterranean it often inhabits waters between 9-20 m depth.

The crab usually occurs just beneath the surface of sand to sandy mud substrates.

GENERAL INFORMATION: females bearing eggs occur from February until August-in North Eastern Atlantic waters and from February until October in the Mediterranean.

The various grooves on the carapace sometimes form a face-like configuration.

The Helmet Crab lives predominantly just below the sand surface. The long pair of second antennae form a tube the tip of which reaches the substrate surface and through which respiratory water is drawn downwards to bathe the gills.

In some regions crabs emerge on the surface at night. Large numbers are sometimes found on beaches after severe storms.

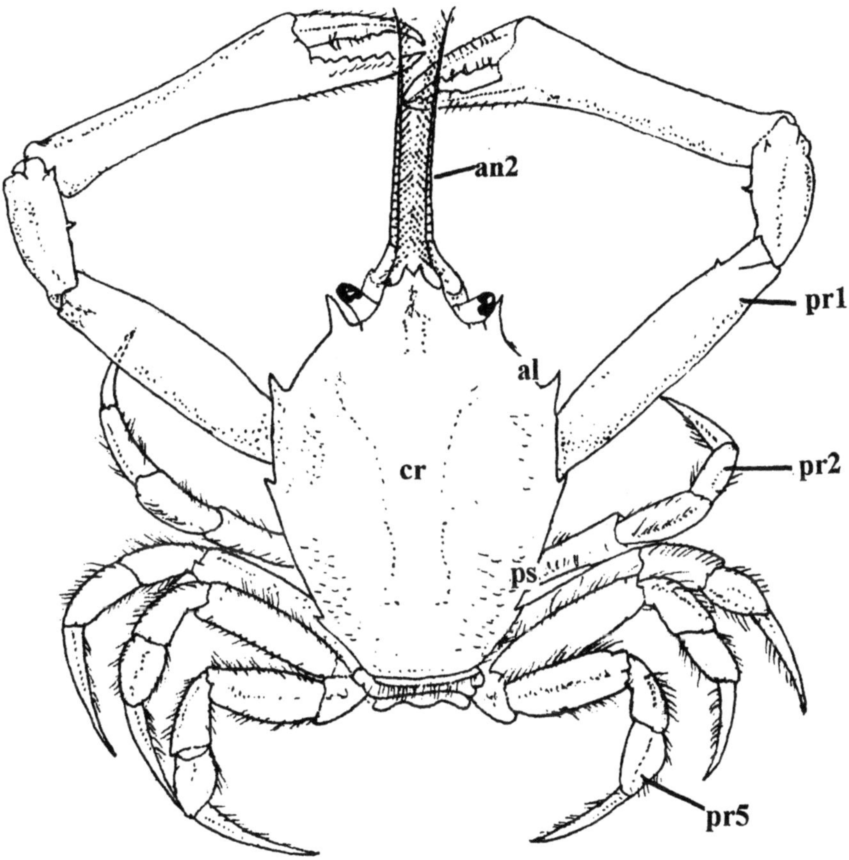

Fig. 10.7 *Corystes cassivelaunus*, Helmet Crab, male. For explanation of annotations see opposite page.

Atelecyclus rotundatus (Olivi, 1792)
Fig. 10.8.
COMMON SYNONYMS: *Atelecyclus septemdentatus.*
DERIVATION: ateles (Gr)-imperfect; kyklos (Gr)-a circle; rotundus (Lt)-rounded.
VERNACULAR NAMES: Circular Crab, Face Crab, Bidenticulate Crab (UK); Cirkelronde Krab (Nd).
DISTINCTIVE FEATURES: carapace (**cr**) almost circular, at most slightly broader than long, dorsal surface with numerous short small transverse carinae; margins with setae; antero-lateral (**al**) margins with nine to eleven sharp teeth; second antennae (**an2**) long; chelipeds (**pr1**) robust and equal in size; second (**pr2**) to fifth (**pr5**) pereiopods stout and with setosed margins.
COLOUR: carapace dorsal surface reddish brown, limbs light brown.
CARAPACE LENGTH: maximum about 4 cm.
DISTRIBUTION: North Eastern Atlantic-from southern Scandinavia southwards to South Africa; also western Mediterranean and Adriatic Sea.

In the North Eastern Atlantic the Circular Crab occurs in waters from about 12 m downwards to over 700 m. In the Mediterranean it is distributed between 20 and 90 m depths.

The crab burrows into sandy and gravel substrates.

GENERAL INFORMATION: when buried this species forms the second antenna into a tube to draw respiratory water from the substrate surface in a similar manner to the Helmet Crab.

In North Eastern Atlantic waters females bear eggs from February until June although breeding probably occurs throughout most of the year.

One of true crab species not infrequently found in large numbers in the stomachs of Cod and Rays.

The only other species occurring in the North Eastern Atlantic and Mediterranean is *Atelecyclus undecimdentatus* that can be recognised by having a considerably broader than long carapace with strongly converging postero-lateral margins.

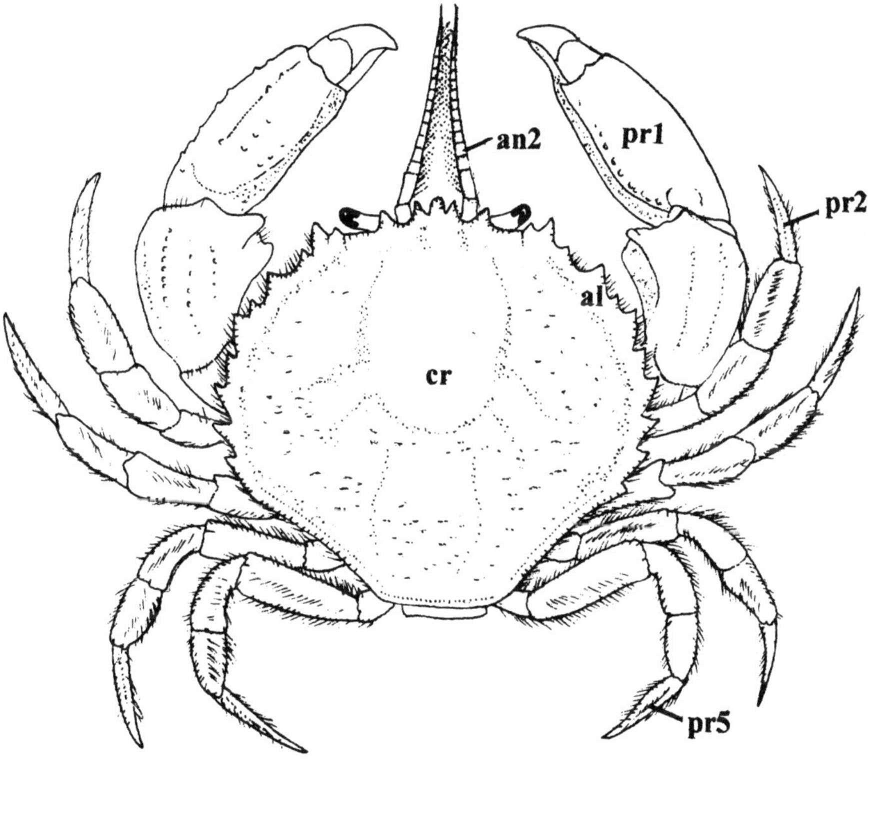

Fig. 10.8 *Atelecyclus rotundatus*, Circular Crab. For explanation of annotations see opposite page.

Cancer pagurus Linnaeus, 1758
Fig. 10.9, Pl. 24.
COMMON SYNONYMS: none.
DERIVATION: cancer (Lt)-a crab; pagouros (Gr)-a crab.
VERNACULAR NAMES: European Edible Crab, Ox Crab, Common Crab, Great Crab, Guernsey Crab (UK); Krabbe (Nw); Krabbtaska (Sw); Taskekrabbe (Dn); Nordsee-Taschenkrebs (Gn); Noordzeekrab (Nd); Tourteau Poupart, Poinclos, Chancre de Drague (Fr); Buey de Mar (Sp); Sapateira (Pt); Granciporro (It); Kavouras (Gr); Asil Pavurya (Tr).
DISTINCTIVE FEATURES: carapace (**cr**) broadly oval; antero-lateral (**al**) margins cut into broad lobes giving 'pie crust' appearance; chelipeds (**pr1**) robust and smooth; second (**pr2**) to fifth (**pr5**) pereiopods stout, distal segments with tufts of short setae.
COLOUR: carapace reddish brown, small crabs purple brown; pereiopods lighter; dactyl and propodal extremities of chelae black.
CARAPACE LENGTH: rarely 16 cm, commonly marketed size 8-9 cm [12-13 cm carapace breadth].
DISTRIBUTION: North Eastern Atlantic-from Scandinavia southwards to Portugal. Mediterranean records probably relate to crabs imported from North Atlantic regions.

Small crabs are often numerous in intertidal region. Large crabs occur offshore downwards to about 90 m on rocky substrates.

GENERAL INFORMATION: females bearing eggs occur throughout the year; eggs are incubated for 7-8 months.

The meat of the European Edible Crab is considered high in value and the crab is extensively fished off North Eastern Atlantic coasts, in particular, the Northumberland and Yorkshire coasts of England, but it is exploited locally in many other regions.

The European Edible Crab is most frequently fished from April until November with baited pots and creels of similar design to those used for clawed lobsters. European Council regulations govern permitted minimum landed sizes, measured as carapace widths. In UK waters these sizes have been set according to region. For example, 15 cm for the Norfolk coast, 16 cm (males) and 14 cm (females) for Devon and Cornwall. Soft-shelled crabs and females with eggs must be returned to the sea. Local regulations are available from the area fisheries inspector.

The only other species of this genus occurring in the North Eastern Atlantic is *Cancer bellianus*, easily recognised by having serrated lobes on the antero-lateral margins of the carapace. This species is occasionally caught in creels set in relatively deep water.

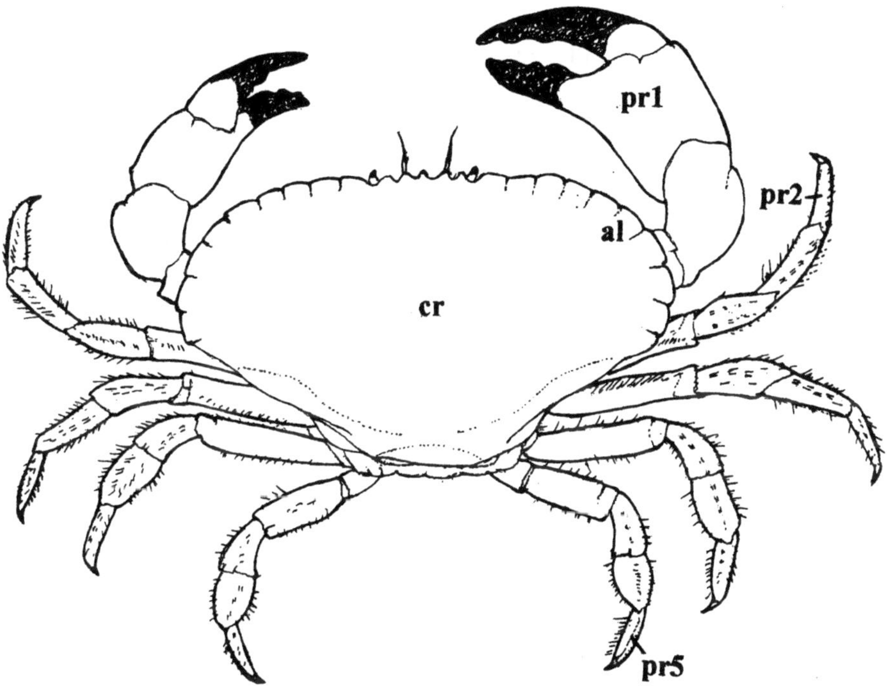

Fig. 10.9 *Cancer pagurus*, European Edible Crab. For explanation of annotations see opposite page.

Chaceon affinis (A. Milne Edwards & Bouvier, 1894)
Fig. 10.10.
COMMON SYNONYMS: *Geryon affinis*
DERIVATION: Chaceon-after the American carcinologist Dr Fenner A Chace Jr; affinis (L)-related to.
VERNACULAR NAMES: Red Geryonid Crab.
DISTINCTIVE FEATURES: carapace (**cr,** fig. A) hexagonal in outline, broader than long, antero-lateral (**al**) margins with five indistinct teeth, dorsal surface of orbital (**ob**) region with a small hump; chelipeds (**pr1**) stout and equal in size, chelae (fig. C) somewhat elongated and relatively straight; second (**pr2**) to fifth (**pr5**) pereiopods long and relatively thin, dactyl (**dcy**) of fifth compressed laterally (ie. from the top surface downwards as shown in fig. B).
COLOUR: carapace and limbs yellowish/orange to red, sometimes with patches of dark brown.
CARAPACE LENGTH: maximum about 15.5 cm.
DISTRIBUTION: North Eastern Atlantic-from Iceland southwards to Cape Verde Islands.

The Red Geryonid Crab occurs on sand to muddy sand and rocky substrates in depths from 130-2047 m.

GENERAL INFORMATION: females with eggs occur in March and August.

Red Geryonid Crabs are occasionally landed by vessels trawling in deep offshore waters.

Although comparable in size with the European Edible Crab (p.136), the Red Geryonid Crab has not been commercially exploited in the North Eastern Atlantic probably because of the difficulties involved in developing a suitable deep water creel fishery and the sparse numbers caught.

For distinctive features between this species and the imported West African Geryonid Crab see p.227.

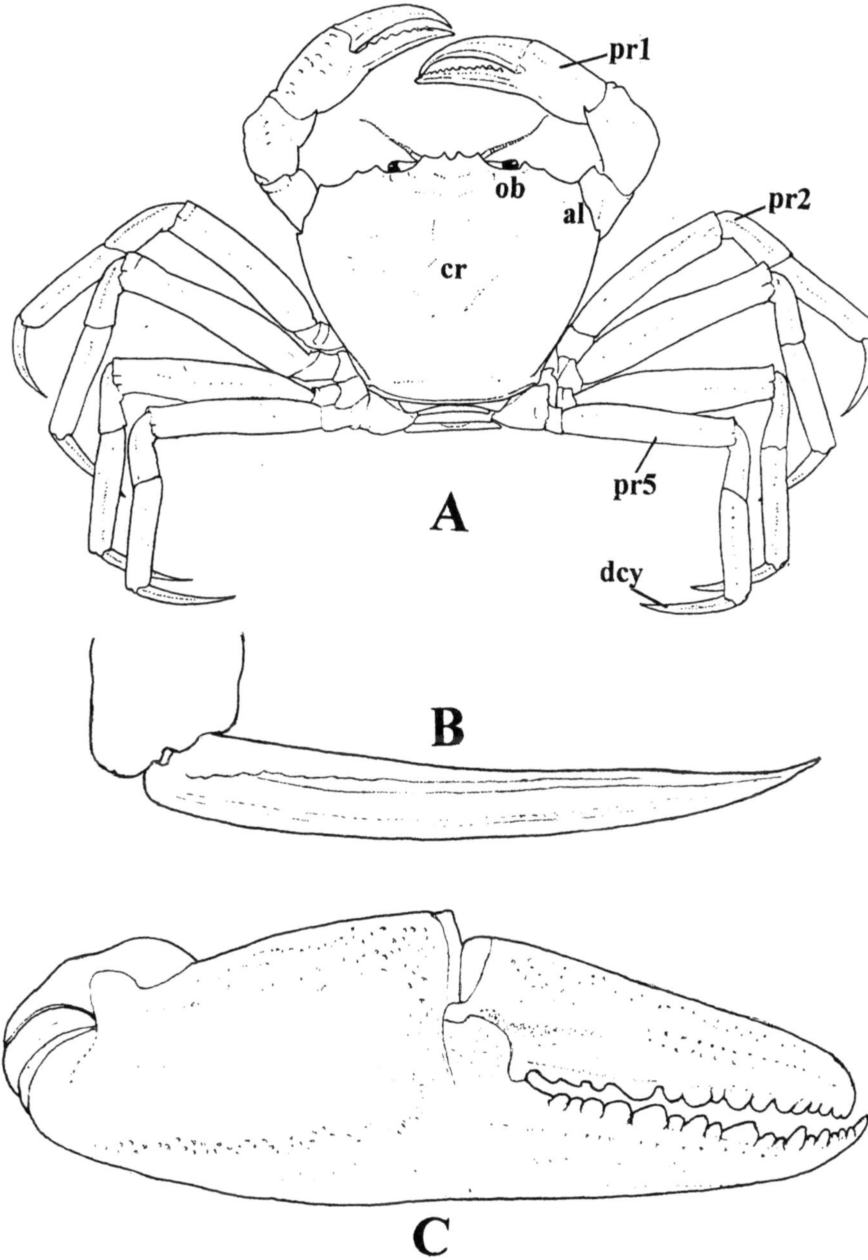

Fig. 10.10 *Chaceon affinis*, Red Geryonid Crab. A. Dorsal aspect of whole crab. B. Dactylus (dcy) of left fifth pereiopod (pr5) from outer aspect. C. Right chela from outer aspect. For explanation of annotations see opposite page.

Polybius henslowii Leach, 1820
Fig. 10.11.
COMMON SYNONYMS: none.
DERIVATION after Polybius-Greek historian or son-in-law of Hippocrates or Secretary to Emperor Claudius; henslowii-after John Steven Henslow Professor of Botany at Cambridge University.
VERNACULAR NAMES: Sardine Swimming Crab, Nipper Crab, Henslow's Swimming Crab (UK); Henslow's Schwimmkrabbe (Gn); Crabe à Sardine, Krank-avel (Fr); Patelo, Pateixo (Sp); Carraça, Pilado (Pt); Yengeç (Tr, perhaps for crabs imported from western Mediterranean).
DISTINCTIVE FEATURES: carapace (**cr**) circular and smooth; antero-lateral (**al**) margins with five broad teeth; chelipeds (**pr1**) usually equal in size, dactyl (**dcy**) noticeably curved and sharp; second (**pr2**) to fifth (**pr5**) pereiopod strongly flattened, dactyls of second to fourth lance-shaped; dactyls of fifth pereiopods paddle shaped.
COLOUR: dorsal surface of carapace reddish brown, sometimes with white markings; distal parts of chelipeds orange yellow.
CARAPACE LENGTH: maximum about 4.4 cm.
DISTRIBUTION: North Eastern Atlantic-from region of Orkney Islands southwards to Morocco. Western Mediterranean eastward to about Malta region.

The Sardine Swimming Crab occurs in off-shore surface waters downwards to depths in excess of 2000 m.

GENERAL INFORMATION: females, that rarely occur, bear eggs from May until December in the North Eastern Atlantic and during spring in the Mediterranean.

The Sardine Swimming Crab swarms in enormous numbers near the sea surface and preys voraciously upon fish shoals (particularly sardines). At these times crabs often damage fishing nets.

When quantities are landed by Portuguese fishermen they are used as land fertilizer.

Large numbers are occasionally cast ashore after severe storms.

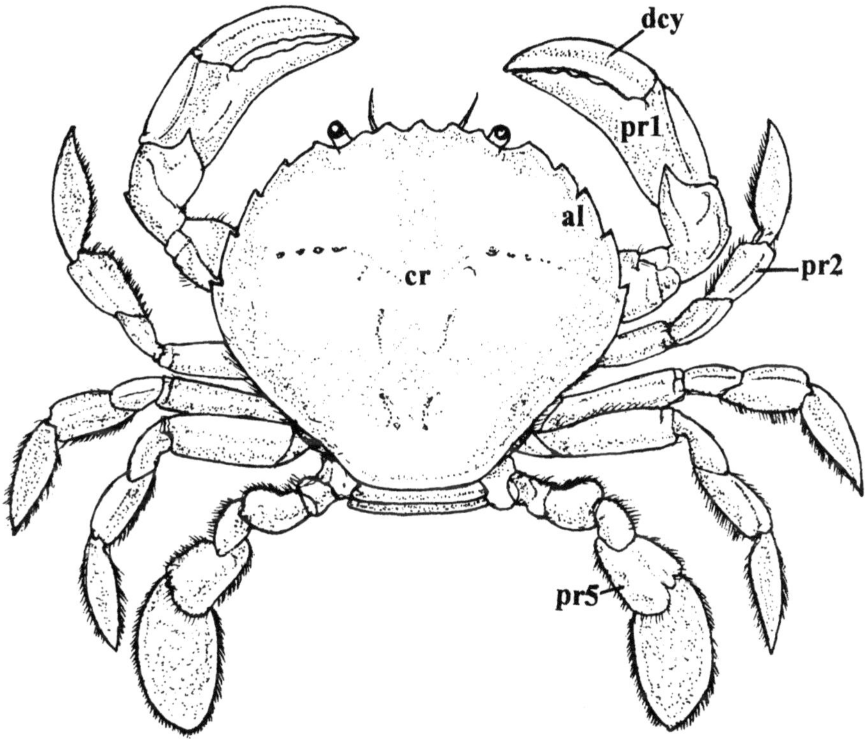

Fig. 10.11 *Polybius henslowii*, Sardine Swimming Crab. For explanation of annotations see opposite page.

Liocarcinus depurator (Linnaeus, 1758)
Fig. 10.12.
COMMON SYNONYMS: *Portunus depurator*; *Macropipus depurator*.
DERIVATION: leio (Gr)-to make smooth; karkinos (Gr)-a crab; depurgo (Lt)-I cleanse, alluding to crab's scavenging habits.
VERNACULAR NAMES: Harbour Swimming Crab, Cleanser Crab, Mary Crab, Blue-legged Swimming Crab (UK); Simkrabba (Sw); Svømmekrabbe (Nw, Dn); Rauhe Schwimmkrabbe (Gn); Blauwpootzwemkrab, Zwemkrab (Nd); Etrille Pattes Bleues (Fr); Falsa Nécora, Jaiba (Sp); Granchio, Masineta, Cavuru (It); Yengeç (Tr).
DISTINCTIVE FEATURES: carapace (**cr**) dorsal surface with numerous small transversely placed setosed ridges; teeth on frontal (**fr**) region very pointed, antero-lateral (**al**) margins with five sharp teeth and setae; chelipeds (**pr1**) equal in size, propodus (**pro**) with conspicuous longitudinal ridges, carpus (**cpr**) with large acute process on inner margin; carpus (**cpr**) and propodus (**pro**) of second (**pr2**) to fourth (**pr4**) pereiopods with longitudinal ridges; fifth pereiopod (**pr5**) paddle-shaped.
COLOUR: carapace and pereiopods reddish brown to mauve; mid-portion of carapace with curved band of white spots; dactylus of fifth pereiopod bright blue.
CARAPACE LENGTH: maximum about 4 cm.
DISTRIBUTION: North Eastern Atlantic-from northern Norway southwards to Spanish Sahara (West Africa). Whole of Mediterranean.

The Harbour Swimming Crab occurs in onshore waters deeper than about 5 m downwards to beyond 300 m on sand, muddy sand and gravel substrates.

GENERAL INFORMATION: females with eggs occur throughout the year. Crabs often inhabit harbours near moored fishing vessels and feed on trash thrown overboard. Sometimes small groups of crabs swim near the surface.

The Harbour Swimming Crab is distinguished from the other nine species of *Liocarcinus* occurring in the North Eastern Atlantic/Mediterranean region by the following combined features. Transverse setosed ridges present on the carapace; very pointed teeth on the carapace frontal region; carapace margins and anterior margins of merus of second to fourth pereiopods not *densely* setosed.

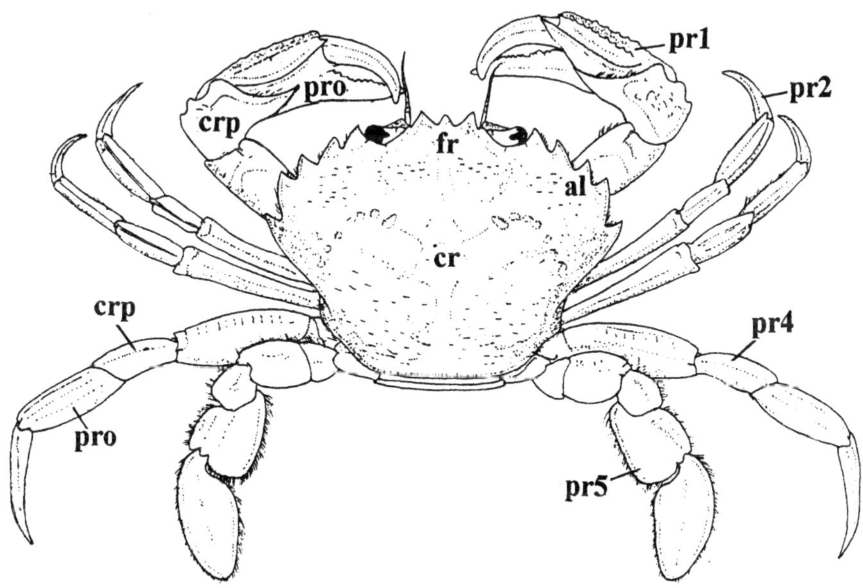

Fig. 10.12 *Liocarcinus depurator*, Harbour Swimming Crab. For explanation of annotations see opposite page.

Liocarcinus holsatus (Fabricius, 1798)
Fig. 10.13.
COMMON SYNONYMS: *Portunus holsatus*; *Macropipus holsatus*.
DERIVATION: leio (Gr)-to make smooth; karkinos (Gr)-a crab; holsatus-named by J. Fabricius after the locality Schleswig-Holstein in North Western Germany from where specimens were presumably collected.
VERNACULAR NAMES: Flying Swimming Crab, Livid Swimming Crab, Flying Gillie (UK); Svømmekrabber (Sc); Zweemkrab, Gewone Zwemkrab (Nd); Crabe Nageur (Fr).
DISTINCTIVE FEATURES: carapace (**cr**) broad, sub-oval and slightly granular; frontal region (**fr**) with sub-acute lobes, antero-lateral (**al**) margins with five setosed teeth, first to fourth very broad and flat (particularly the fourth), fifth acute; postero-lateral (**ps**) margins strongly convergent; chelipeds (**pr1**) equal in size, inner margin of carpus (**crp**) with a noticeably stout acute process, outer margin with two processes; carpus (**crp**) to propodus (**pro**) of second (**pr2**) to fourth (**pr4**) pereiopods with longitudinal ridges; fifth pereiopod (**pr5**) paddle-shaped, merus (**mrs**) short, dactylus (**dcy**) broad.
COLOUR: dorsal surface of carapace and limbs greenish brown; mid-portion of carapace often with a curved band of white spots.
CARAPACE LENGTH: maximum about 3.9 cm.
DISTRIBUTION: North Eastern Atlantic-from northern Norway southwards to southern Portugal. Absent from Mediterranean.

The Flying Swimming Crab occurs rarely in shallows of the lowest shore region and downwards to about 400 m on variable substrates from muddy sand to gravel.

GENERAL INFORMATION: females with eggs occur from April until September.

This crab is occasionally abundant near or just within estuaries.

The combined features of the sub-acute frontal teeth of the carapace, the noticeably flattened, broad pair of fourth antero-lateral teeth and the two processes on the outer margin of the carpus distinguish the Flying Swimming Crab from the other nine *Liocarcinus* species occurring in the North Eastern Atlantic.

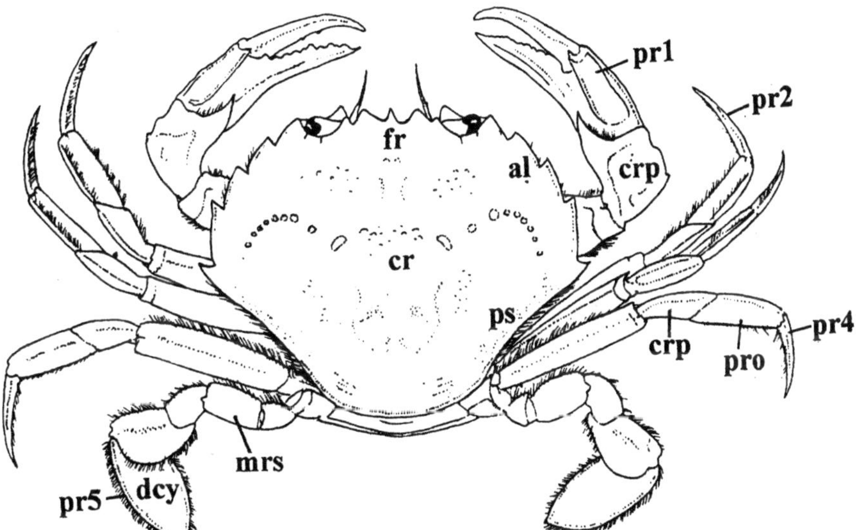

Fig. 10.13 *Liocarcinus holsatus*, Flying Swimming Crab. For explanation of annotations see opposite page.

Necora puber (Linnaeus,1767)
Fig.10.14, Pl. 25.
COMMON SYNONYMS: *Portunus puber*; *Macropipus puber*; *Liocarcinus puber*.
DERIVATION: necora-vernacular name for this species in Galicia, Spain; puber (Lt)-hairiness, denoting fine pubescence on carapace surface.
VERNACULAR NAMES: Velvet Swimming Crab, Velvet Fiddler Crab, Kerry Witch (UK); Simkrabba (Sw); Svømmekrabbe (Nw, Dn); Samtkrabbe (Gn); Fluwele Zwemkrab (Nd); Etrille Commune, Crabe à Laine, Demoiselle (Fr); Nécora, Anacra (Sp); Navalheira (Pt); Siderokavouras (Gr); Çalpara (Tr).
DISTINCTIVE FEATURES: carapace (**cr**) dorsal surface covered with fine pubescence; margin of frontal (**fr**) region cut into seven to ten unequal teeth; antero-lateral (**al**) margins with five sharp teeth; chelipeds (**pr1**) equal in size and stout, propodus (**pro**) with longitudinal tuberculate ridges; carpus (**crp**) and propodus (**pro**) of second (**pr2**) to fifth (**pr5**) pereiopods with conspicuous ridges; dactylus (**dcy**) of fifth with a prominent medial ridge.
COLOUR: carapace brown to black; parts of pereiopods bright blue spotted with red.
CARAPACE LENGTH: maximum about 7.8 cm.
DISTRIBUTION: North Eastern Atlantic-from southern Norway southwards to Spanish Sahara (N.W. Africa). Whole of Mediterranean but much less common, also Black Sea.

The Velvet Swimming Crab occurs from lower shore downwards to about 80 m depth chiefly on rock and stone substrates.

GENERAL INFORMATION: females with eggs occur from February until November.

A very aggressive crab. In the North Eastern Atlantic it is often the next commonest species to the North Atlantic Shore Crab.

Sold in European markets for food. The meat is chiefly in the chelipeds.

The Velvet Swimming Crab is easily distinguished from the other nine species of Swimming Crabs occurring in the North Eastern Atlantic/Mediterranean region by the numerous small teeth on the frontal margin of the carapace and the fine pubescence on the carapace dorsal surface.

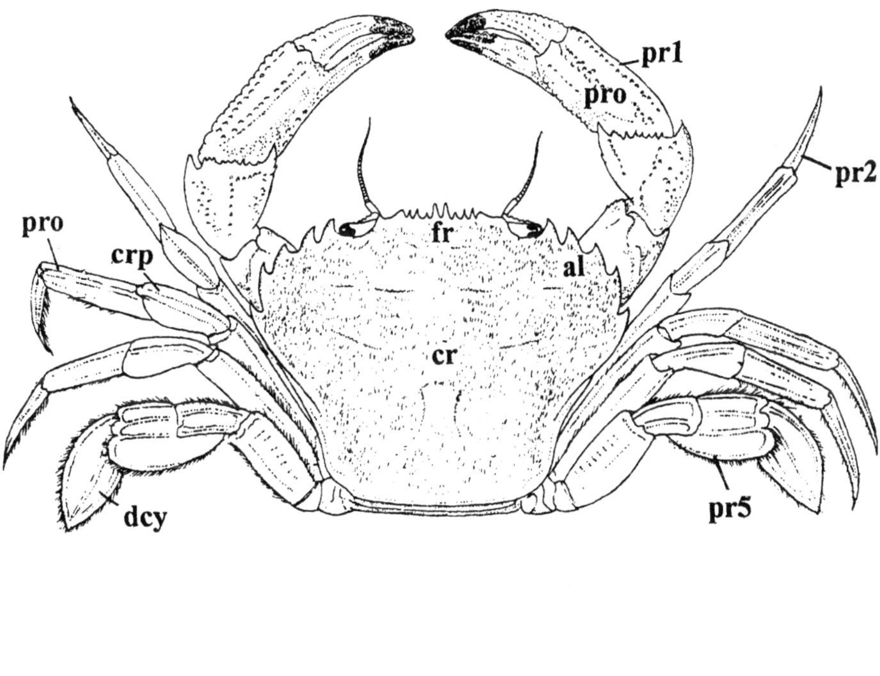

Fig. 10.14 *Necora puber*, Velvet Swimming Crab. For explanation of annotations see opposite page.

Callinectes sapidus Rathbun, 1896
Fig. 10.15, Pl. 26.
COMMON SYNONYMS: none.
DERIVATION: kalos (Gr)-beautiful; nektes (Gr)-swimmer; sapidus (Lt)-savoury.
VERNACULAR NAMES: Chesapeake Bay Swimming Crab, Blue Crab (UK); Blakrabba (Sw); Sinitaskurapu (Fi); Blakrabbe (Nw); Blaukrabbe (Gn); Blauwe Zwemkrab (Nd); Crabe Bleu (Fr); Cangrejo Azul (Sp); Navalheira-azul (Pt); Granchio Nuotatore (It); Italikos Kavouras (Gr); Mavi Yengeç (Tr).
DISTINCTIVE FEATURES: carapace (**cr**) very broad; antero-lateral (**al**) margins with nine sharp teeth, most posterior pair stout and very long; chelipeds (**pr1**) stout and equal in size, inner margin of merus (**mrs**) with three stout spines; second (**pr2**) to fifth (**pr5**) pereiopods flattened; dactyl (**dcy**) of fifth broadly sub-oval.
COLOUR: dorsal surface of carapace greyish to bluish green, spines with red tips; distal parts of chelae blue in male, red in female.
CARAPACE LENGTH: maximum about 8.5 cm.
DISTRIBUTION: a native of the Eastern American Atlantic coast and distributed from Nova Scotia southwards to Uruguay.

The Chesapeake Bay Swimming Crab occurs sporadically in North Eastern Atlantic waters (Denmark, eastern British Isles, Netherlands, French coasts). It is well established in the eastern Mediterranean.

The crab inhabits estuaries and shallow onshore waters. It lives on and burrows into various substrates but often occurs on muddy sea beds.
GENERAL INFORMATION: the Chesapeake Bay Swimming Crab is the most important species commercially fished in the USA. Soft-shelled crabs 'peelers'(a UK term) are often seen in restaurants in the USA.

Crabs sold in Mediterranean fish markets are probably locally caught. Whole crabs, fresh and frozen meat seen in UK markets is usually imported from the USA. The meat is highly esteemed.

This species is easily distinguished from all other species of swimming crabs occurring in the North Eastern Atlantic/Mediterranean regions (or imported species) by the very broad carapace and large posterior pair of spines on the carapace antero-lateral margin and a prominent undivided medial frontal process.

In the author's opinion the vernacular name 'Blue Crab' should be restricted to the Indo-Pacific species *Portunus pelagicus* that is occasionally imported into european markets.

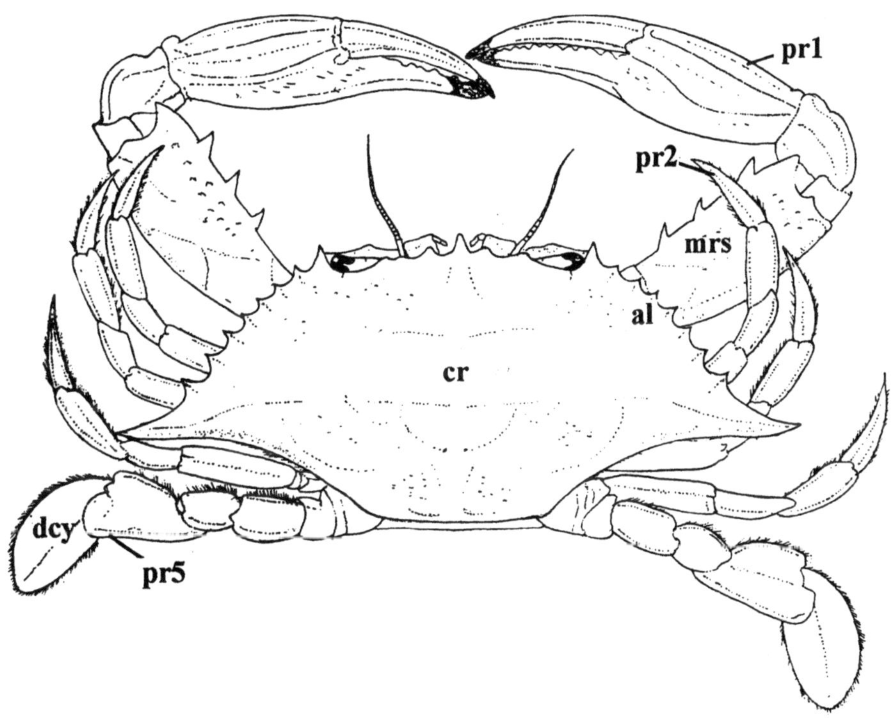

Fig. 10.15 *Callinectes sapidus*, Chesapeake Bay Swimming Crab. For explanation of annotations see opposite page.

Carcinus maenas (Linnaeus, 1758)
Fig.10.16.
COMMON SYNONYMS: *Carcinides maenas.*
DERIVATION: karkinos (Gr)-a crab; maenas (Lt)-either derived from Maenadis (Lt) priestesses who worshiped at the altar of Priapus, or more probably from mainas (Gr)-enraged, frantic, and referring to the aggressive behaviour of this crab.
VERNACULAR NAMES: North Atlantic Shore Crab, Harbour Crab, Green Crab (UK); Rantataskurapu (Fi); Strandkrabbe (Sc & Gn); Strandkrab (Nd); Crabe Vert, Crabe Enragé, Chancre (Fr); Cangrejo Verde, Caranguexo (Sp); Cangrexo (Pt); Kavouras (Gr); Cingene pavuryasi (Tr). (The names used in Greece and Turkey may refer to imported European *Carcinus maenas).*
DISTINCTIVE FEATURES: carapace (**cr**) broad and sub-oval; antero-lateral margins (**al**) with five sharp teeth; chelipeds (**pr1**) slightly unequal in size; second (**pr2**) to fifth (**pr5**) pereiopods stout, dactyls (**dcy**) curved and sharp distally, those of fifth pereiopods fringed with setae, prominent in young crabs.
COLOUR: variable, dorsal surface of carapace commonly dark green to grey, often orange in offshore crabs. Small intertidal crabs with white to black markings; pereiopods of most crabs brown to greenish yellow or orange.
CARAPACE LENGTH: maximum about 6 cm.
DISTRIBUTION: North Eastern Atlantic-from northern Norway southwards to Mauritania (West Africa). Also introduced to Atlantic coast of USA, Red Sea, Madagascar, Sri Lanka and Australia.

The North Atlantic Shore Crab occurs intertidally downwards to about 30 m on all types of substrates. Not uncommon in estuaries.

GENERAL INFORMATION: the commonest intertidal crab. Probably breeds throughout year.

This crab is occasionally offered for sale in a few European markets but not seriously commercially exploited. Used as bait by anglers.

The Mediterranean Shore Crab, *Carcinus aestuari* is distinguished by having a smoother and more raised carapace, more pointed antero-lateral teeth and the external margin of the cheliped carpus with setae. This species occurs from the Canary Islands, throughout the Mediterranean and Black Sea.

Vernacular names: Mittelmeer-Strandkrabbe (Gn); Crabe de Méditerranée (Fr); Cangrejo Mediterráneo (Sp); Caranguejo Mediterráneo (Pt); Granchio Ripario (It); Ariokavouras (Gr); Caganoz (Tr).

The Venetians eat this species when soft and newly moulted (termed 'moleche').

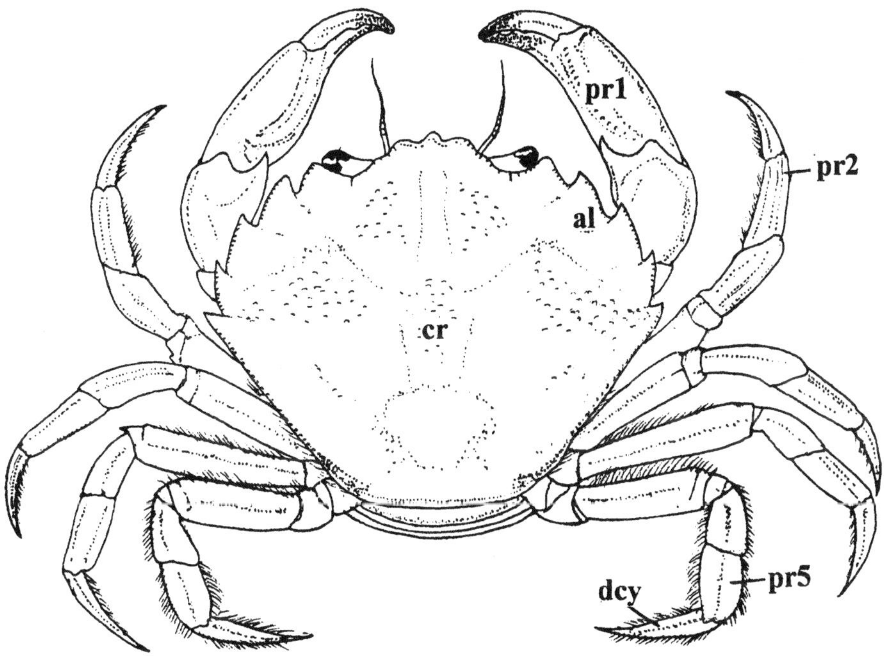

Fig. 10.16 *Carcinus maenas*, North Atlantic Shore Crab. For explanation of annotations see opposite page.

Eriphia verrucosa (Forskål, 1775)
Fig. 10.17, Pl. 29.
COMMON SYNONYMS: *Eriphia spinifrons.*
DERIVATION: it is difficult to comprehend what Pierre Latreille had in mind when he established the generic name for this crab. Eriphia (Lt)-an unknown plant but more likely eriphos (Gr)-a small goat, perhaps alluding to the crab's antennae resembling small horns; verruca (Lt)-a wart, alluding to its warty appearance.
VERNACULAR NAMES: Warty Xanthid Crab (UK); Italienischer Taschenkrebs (Gn); Crabe Verruqueux (Fr); Cangrejo Moruno, Boron (Sp); Caranguejo-mouro (Pt); Favollo, Granzoporo, Paura (It); Petrokavouras (Gr); Pavurya (Tr).
DISTINCTIVE FEATURES: carapace (**cr**) dorsal surface with transverse granular ridges; antero-lateral (**al**) margins with six to seven sharp spines decreasing in size posteriorly and many bearing distal spinules; frontal (**fr**) region with two transverse rows of spines; chelipeds (**pr1**) very robust, right usually slightly larger than left, with tubercles arranged in longitudinal rows, distal parts of chelipeds black; second (**pr2**) to fifth (**pr5**) pereiopods stout and very setosed, dactylus (**dcy**) sharp pointed.
COLOUR: overall brownish red to brownish green; carapace sometimes with yellow patches or spots.
CARAPACE LENGTH: maximum about 9 cm.
DISTRIBUTION: North Eastern Atlantic-from southern Brittany (France) southwards to Azores. Whole of Mediterranean including Black Sea.

The Warty Xanthid Crab occurs on rocky substrates and among algae from the lowest part of the shore downwards to about 10 m depth and often in regions where waves break on rocks.

GENERAL INFORMATION: in the Mediterranean females bearing eggs occur during May, July and August.

This crab is fished and eaten in the Mediterranean; it is sometimes seen in fish markets.

Adults of the Warty Crab are distinguished from other xanthid crabs occurring in the North Eastern Atlantic/Mediterranean region by the large tuberculate chelipeds and spinules on the carapace antero-lateral spines.

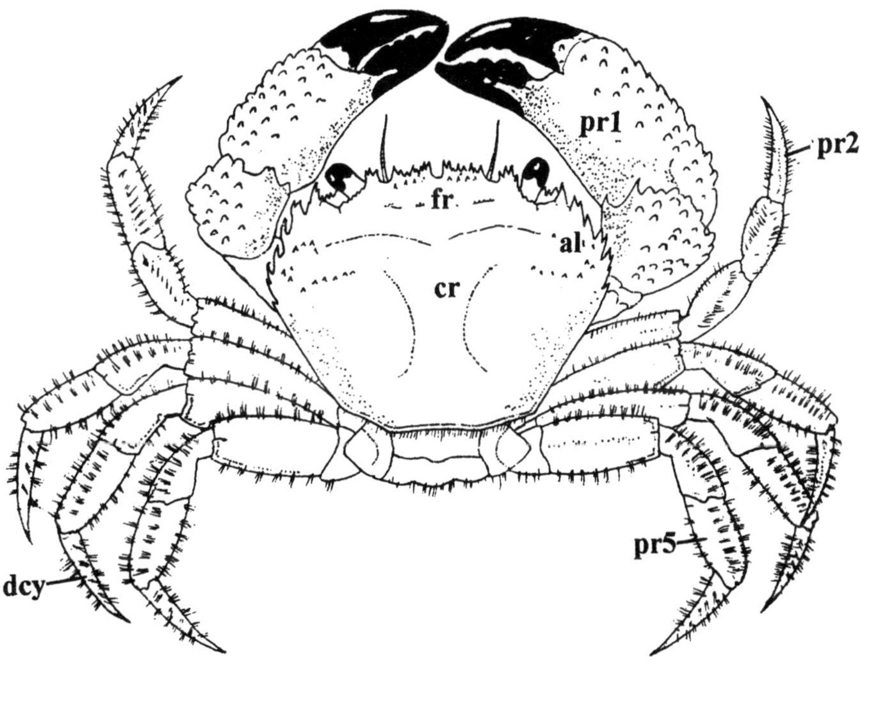

Fig. 10.17 *Eriphia verrucosa*, Warty Xanthid Crab. For explanation of annotations see opposite page.

Xantho incisus Leach, 1814
Fig. 10.18.
COMMON SYNONYMS: *Xantho floridus*.
DERIVATION: xanthos (Gr)-various shades of yellow, alluding to colour of crab; incido (Lt)-cut into, presumably referring to the carapace antero-lateral margins that are cut into lobes.
VERNACULAR NAMES: Furrowed Xanthid Crab, Montagu's Crab (UK).
DISTINCTIVE FEATURES: dorsal surface of carapace (**cr**) finely granular (almost smooth to unaided eye); antero-lateral (**al**) margins with four very rounded lobes, fourth smallest; chelipeds (**pr1**) equal in size and very robust; second (**pr2**) to fifth (**pr5**) pereiopods stout and relatively short, upper margins of carpus (**crp**) of these limbs smooth (at least in the proximal half).
COLOUR: carapace yellowish-brown, sometimes with patches of deeper brown; dactyls and distal part of chelipeds deep black.
CARAPACE LENGTH: maximum about 2.5 cm.
DISTRIBUTION: North Eastern Atlantic-from west coast of Scotland southwards to Gulf of Guinea (West Africa). Probably absent from North Sea and adjacent coastal regions.

The Furrowed Xanthid Crab occurs from the lower shore downwards to about 100 m depth, often on stony substrates.
GENERAL INFORMATION: females bearing eggs occur from March through to August.

Many authorities recognise two subspecies. The Northern Atlantic *Xantho incisus incisus* and the Mediterranean *Xantho incisus granulicarpus*. The latter has a more granular carapace and the carpal upper margins of the second to fifth pereiopods are irregular and strongly granular. However, forms intermediate between these two subspecies also occur that suggests doubt about the validity of their subspecific status.

The Furrowed Xanthid Crab is sometimes seen in public aquaria.

The closely related Less Furrowed Xanthid Crab (*Xantho pilipes*) can be recognised by the presence of setae fringing the posterior pairs of carapace antero-lateral teeth that are sub-acute; there are also long setae on the margins of the fourth and fifth pereiopods.

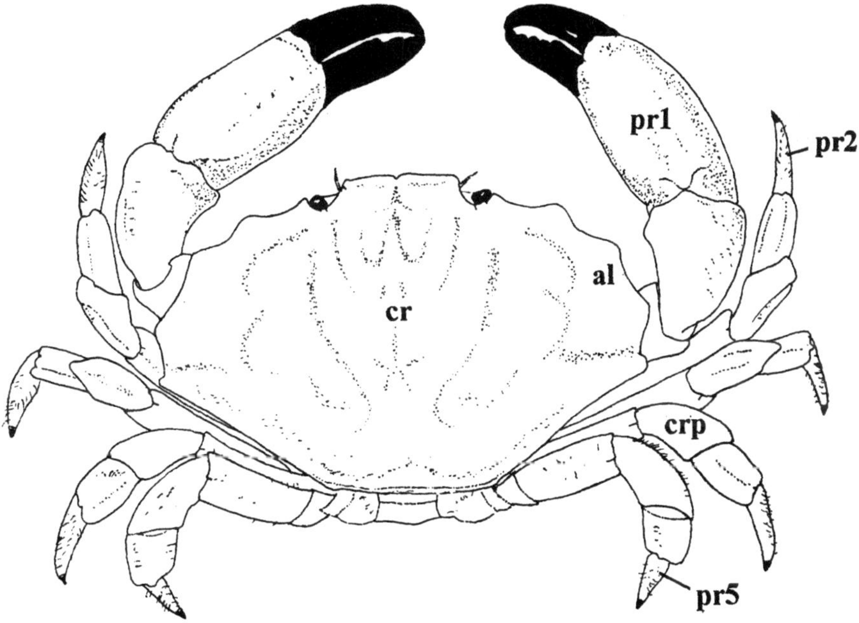

Fig. 10.18 *Xantho incisus*, Furrowed Xanthid Crab. For explanation of annotations see opposite page.

Pilumnus hirtellus (Linnaeus, 1761)
Fig. 10.19.
COMMON SYNONYMS: none.
DERIVATION: Pilumnus (Lt)-Roman deity, personification of the pestle; specific name possibly derived from hirt (Lt)-hairy and tellus (Lt)-the Earth, perhaps alluding to the often mud-covered hairy body.
VERNACULAR NAMES: Bristly Xanthid Crab, Hairy Crab (UK); Ruig Krabbetje (Nd); Crabe Poilu (Fr); Cangrejo Aterciopelado (Sp).
DISTINCTIVE FEATURES: dorsal surface of carapace (**cr**) covered with club-shaped setae, usually dense and sometimes matted with mud; antero-lateral (**al**) margins of carapace with five pointed spines; chelipeds (**pr1**) robust and unequal, propodus (**pro**) with tubercles; second (**pr2**) to fifth (**pr5**) pereiopods relatively stout, dactyls (**dcy**) curved distally and acutely pointed; surfaces of limbs covered with numerous club-shaped setae; male chelipeds noticeably robust.
COLOUR: carapace reddish brown, sometimes purple; outer surfaces of chelipeds brownish to purple, inner surfaces orange to cream coloured; pereiopods sometimes with alternate bands of purple and greyish white.
CARAPACE LENGTH: maximum about 1.5 cm.
DISTRIBUTION: North Eastern Atlantic-from southern Norway southwards to Cape Verde Islands. Whole of Mediterranean and also Black Sea.

The Bristly Xanthid Crab occurs on the lower shore downwards to about 80 m on various substrates from muddy sand to stones and rocks and sometimes in seaweed holdfasts.

GENERAL INFORMATION: females bear eggs from March until November in the North Eastern Atlantic and from March until July and September in the Mediterranean.

Juveniles are occasionally associated with the Oplet anemone (*Anemonia viridis*).

Not infrequently seen in public aquaria.

Six species of this genus occur in the North Eastern Atlantic/Mediterranean regions and some of these are not easily distinguished one from the other. The five antero-lateral teeth on the carapace, the dense covering of setae on the carapace dorsal surface and the relatively stout second to fifth pereiopods are combined features by which the Bristly Crab can be recognised.

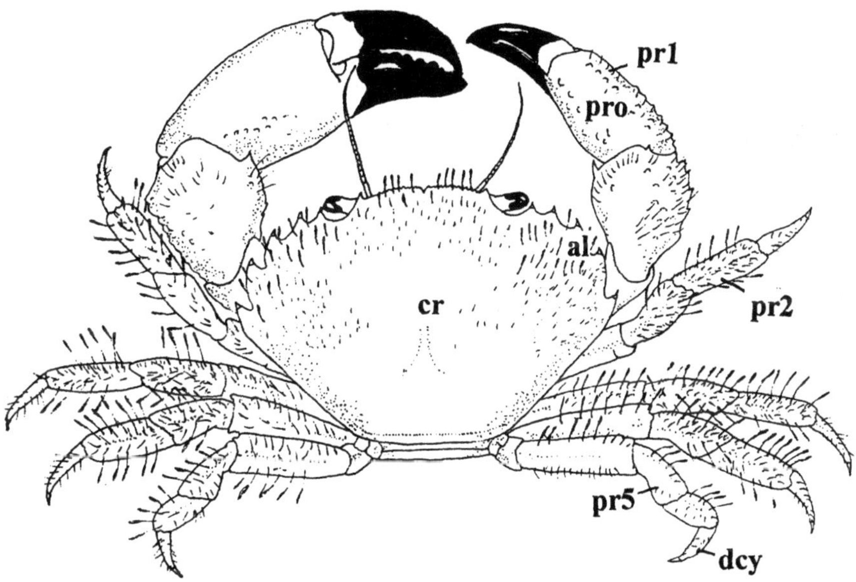

Fig. 10.19 *Pilumnus hirtellus*, Bristly Xanthid Crab. For explanation of annotations see opposite page.

Pinnotheres pisum (Linnaeus, 1767)
Fig. 10.20.
COMMON SYNONYMS: none.
DERIVATION: pinnotheres (Gr)-the crab that lives within and guards the Pinna's shell; pisum (Lt)-a pea, alluding to shape of crab's body.
VERNACULAR NAMES: Linnaeus's Pea Crab (UK); Ertekrabbe, Aetekrabbe (Sc); Erwtenkrabbetje, Mosseelkrabbetje (Nd); Pinnothère (Fr); Cangrejo de los Mejillones (Sp).
DISTINCTIVE FEATURES: dorsal surface of carapace (**cr**) smooth, soft and more or less polygonal in female (fig. A); hard and almost circular in male (fig. B) and with frontal region (**fr**) projecting; eyes and second antennae (**an2**) very small; chelipeds (**pr1**) equal in size, elongated in female (fig. A), noticeably robust in male (fig, B); second (**pr2**) to fifth (**pr5**) pereiopods thin and setosed, dactyls (**dcy**) short and curved.
COLOUR: female-dorsal surface of carapace translucent beneath which is often visible the yellowish organs and red gonads; male-carapace yellowish grey with patches of brown.
CARAPACE LENGTH: maximums-female 1.4 cm; male 8 mm.
DISTRIBUTION: North Eastern Atlantic-from southern Scandinavia southwards to Mauritania (West Africa). Whole of Mediterranean.

Linnaeus's Pea Crab occurs (within bivalve molluscs and ascidians) from the intertidal region to about 150 m depth.
GENERAL INFORMATION: females bearing eggs occur in North Eastern Atlantic waters from April until October. Males swim out of their hosts (after developing swimming setae on the limbs), seek out and mate with females.

This Pea Crab inhabits the mantle cavity of a considerable number of bivalve mollusc species and is therefore sometimes discovered within such shellfish purchased for consumption.

Opinions differ about the taxonomic status of some of the North Eastern Atlantic/Mediterranean species of Pea Crabs. Some authorities recognise two subspecies of *Pinnotheres pisum* (*P. pisum pisum* and *P. pisum pectunculi*). Further research may reveal that many of the 'varieties' mentioned by early naturalists may be valid species or subspecies.

The related Pinnae Pea crab (*Nepinnotheres pinnotheres*) has longer dactyls to the second to fourth pereiopods and the frontal region of the male is not strongly produced.

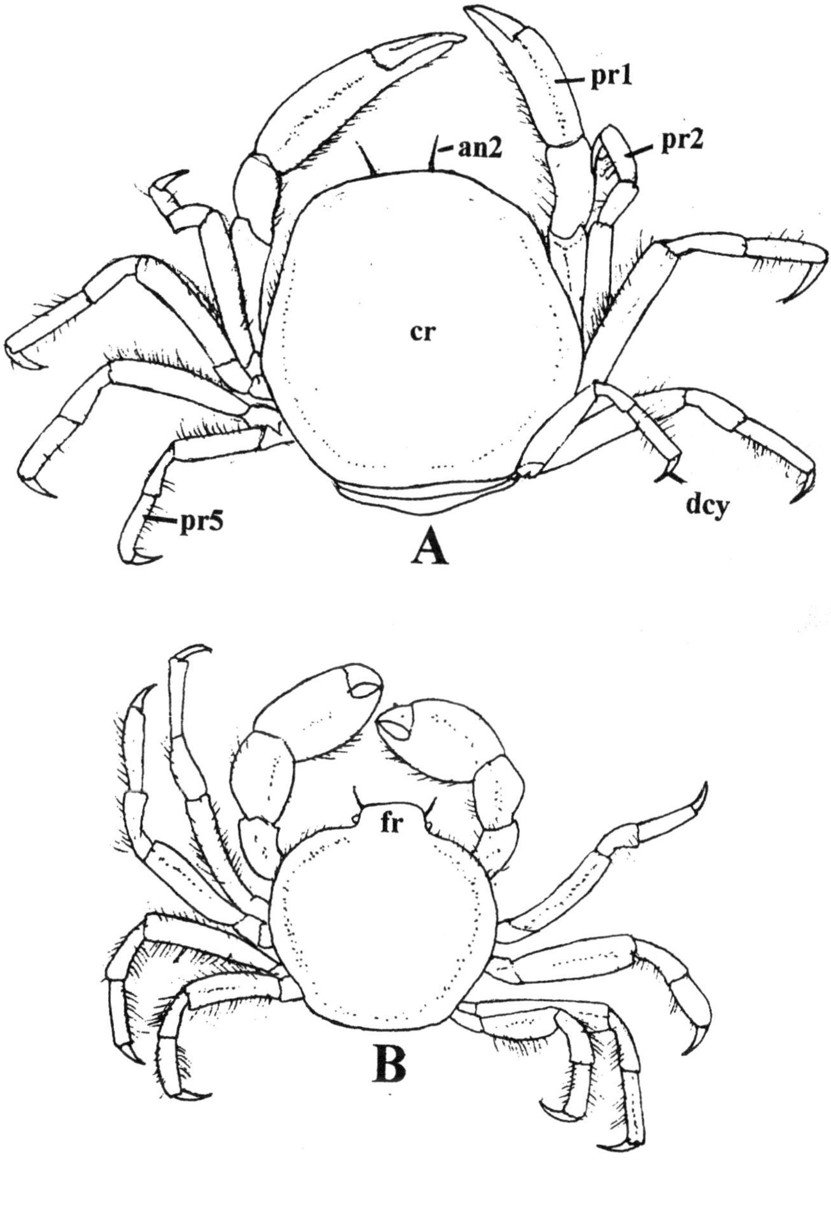

Fig. 10.20 *Pinnotheres pisum*, Linnaeus's Pea Crab. A. Female. B. Male. For explanation of annotations see opposite page.

Pachygrapsus marmoratus (Fabricius, 1787)
Fig. 10.21.
COMMON SYNONYMS: none.
DERIVATION: pachy (Gr)-thick; grapsaios (Gr)-a crab; marmoratus (Lt)-marbled, alluding to colour pattern of crab.
VERNACULAR NAMES: Marbled Rock Crab, Rock Crab (UK).
DISTINCTIVE FEATURES: carapace (**cr**) squarish with many prominent transverse ridges; frontal (**fr**) region very broad; antero-lateral (**al**) margins with three acute teeth, most posterior pair smallest, sometimes very small; chelipeds (**pr 1**) slightly unequal in size, (more robust in male than in female), distal margin of merus (**mrs**) with prominent serration; second (**pr2**) to fifth (**pr5**) pereiopods flattened, broad and stout, margins of propodus (**pro**) and dactylus (**dcy**) with spines.
COLOUR: dorsal surface of carapace varying from violet brown to almost black and often with marbled pattern of yellowish brown; limbs lighter in colour.
CARAPACE LENGTH: maximum about 3.6 cm.
DISTRIBUTION: North Eastern Atlantic-from North Eastern Brittany (France) southwards to Canary Islands. Whole of Mediterranean including Black Sea.

The Marbled Rock Crab occurs on rocks of upper to middle shore, also in crevices of breakwaters and pier piles or similar habitats.
GENERAL INFORMATION: females bear eggs from July until August in the North Eastern Atlantic and from May until July in parts of the Mediterranean.

This crab is often common in intertidal regions of Portuguese, Spanish Atlantic and coastal regions of the Mediterranean.

A very quick moving crab exceedingly difficult to catch on account of its singular ability to slip into the narrowest of crevices.

Distinguished from the two other species of *Pachygrapsus* occurring in the Mediterranean by having three teeth on the carapace antero-lateral margins.

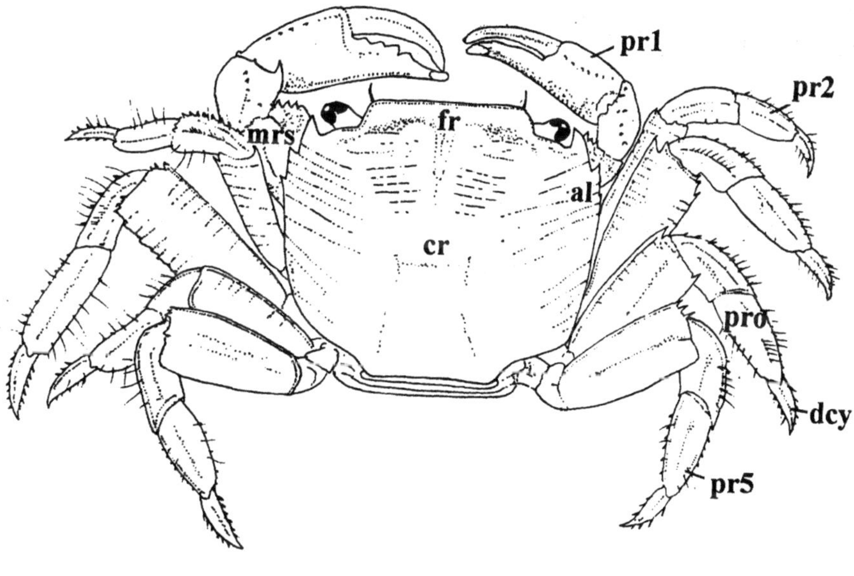

Fig. 10.21 *Pachygrapsus marmoratus*, Marbled Rock Crab. For explanation of annotations see opposite page.

Eriocheir sinensis H.Milne Edwards, 1854.
Fig. 10.22, Pl. 30.
COMMON SYNONYMS: none.
DERIVATION: erion (Gr)-wool; cheir (Gr)-a hand; sinensis (Ni)-of China.
VERNACULAR NAMES: Mitten Crab, Chinese Mitten Crab, Hairy-fisted Crab (UK); Wolhandkrabben (Gn); Chinese Wolhandkrab (Nd); Crabe Chinois (Fr).
DISTINCTIVE FEATURES: carapace (**cr**) squarish with some short transverse ridges; antero-lateral (**al**) margins with three well developed sharp teeth, postero-lateral (**ps**) margins with one small tooth; cheliped (**pr1**) propodal (**pro**) surface covered with conspicuous mat of hair (sometimes patchy in large crabs); second (**pr2**) to fifth (**pr5**) pereiopods stout and flattened, dactyls (**dcy**) curved.
COLOUR: carapace olive green, sometimes with two pairs of distinct white spots; limbs lighter in colour.
DISTRIBUTION: a native of Eastern Asia. Introduced into Europe early this century and now widely distributed from Scandinavia southwards (including UK) to some countries bordering the Mediterranean.
GENERAL INFORMATION: Mitten Crabs live for most of their life in freshwater, usually in upper reaches of rivers. To breed they migrate, often in enormous numbers, downstream into estuaries. When migrating they are sometimes a major pest, piling up against obstructions (eg. sluices) and even leave the water to wander into streets and houses. While in estuaries they damage fishing nets, cutting them to obtain netted fish. Their numerous burrows also cause erosion of river banks.

Raw Mitten Crab is a prized gastronomic delicacy among Chinese and Japanese communities and crabs are sometimes seen in European and UK oriental markets. However, the crab is an intermediate host of lung flukes (*Paragonimus*) and should be well cooked before eating, particularly crabs imported from the Far East.

The Mitten Crab is easily distinguished from all other crab species by the conspicuous mat of hair on the chelipeds.

This species may prove to be synonymous with the Japanese Mitten Crab (*Eriocheir japonica*).

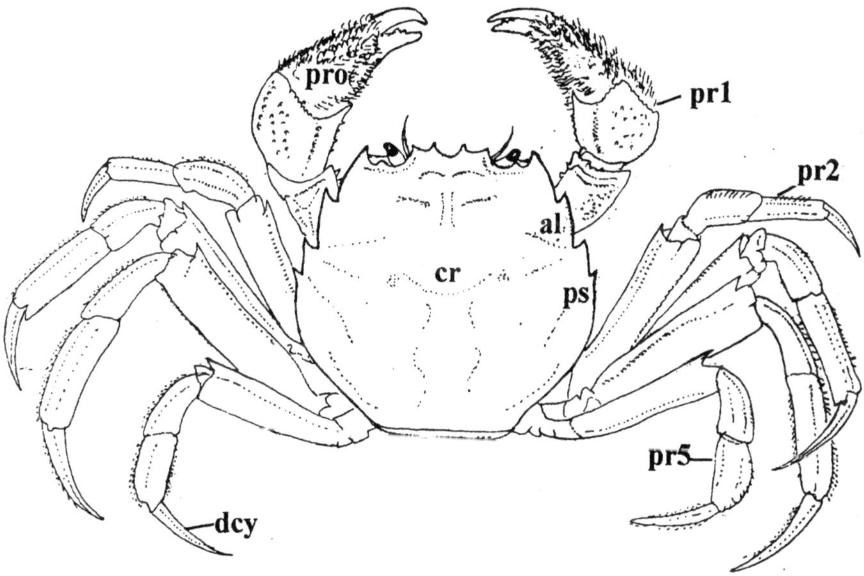

Fig. 10.22 *Eriocheir sinensis*, Mitten Crab. For explanation of annotations see opposite page.

Uca tangeri (Eydoux, 1835)
Fig. 10.23, Pl. 31.
COMMOM SYNONYMS: *Gelasimus tangeri.*
DERIVATION: Uca (Ti)-a fiddler crab; tangeri-name derived from Gulf of Tangiers (Morocco), the locality from where the species was first described.
VERNACULAR NAMES: Moroccan Fiddler Crab, West African Fiddler Crab (UK).
DISTINCTIVE FEATURES: carapace (**cr**) dorsal surface with numerous tubercles; antero-lateral (**al**) and postero-lateral (**ps**) margins strongly converging posteriorly; frontal (**fr**) region with very granular margins on each side of median lobe; eye-stalks very long; left or right cheliped (**pr1**) of male very large, distal part of propodus (**pro**) and dactylus (**dcy**) elongated; chelipeds of female small and usually equal in size; second (**pr2**) to fifth (**pr5**) pereiopods stout, flattened and granular, margins setosed.
COLOUR: carapace violet to almost black, limbs yellowish brown; carapace of female sometimes yellowish.
CARAPACE LENGTH: maximum about 2.5 cm.
DISTRIBUTION: North Eastern Atlantic-from southern Portugal southwards to Angola (West Africa).

The Moroccan Fiddler Crab occurs on sandy to muddy intertidal beaches in which it constructs burrows.

GENERAL INFORMATION: moderate numbers are often seen on intertidal beaches in some regions. Their presence is easily recognised by the males waving the large claw (similar to the motions of the bow arm of a fiddle player) to indicate territorial rights and also to attract and entice females into their burrows for mating.

The large claws of male Moroccan Fiddler Crabs are sometimes offered for sale in Portuguese and Spanish markets for human consumption.

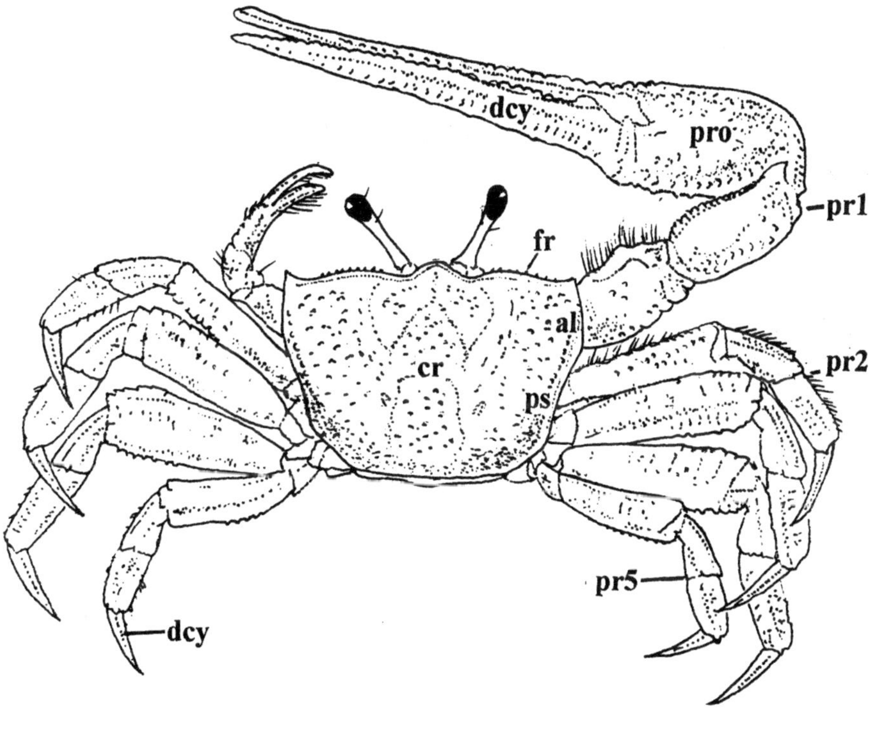

Fig. 10.23 *Uca tangeri*, Moroccan Fiddler Crab, male. For explanation of annotations see opposite page.

Potamon fluviatile fluviatile (Herbst, 1785)
Fig. 10.24.
COMMON SYNONYMS: none.
DERIVATION: potamos (Gr)-a river; fluviatilis (Lt)-of or belonging to a river.
VERNACULAR NAMES: none in common use but Italian Freshwater Crab is proposed here, denoting the country from where this subspecies was first described.
DISTINCTIVE FEATURES carapace (**cr**, fig. A) sub-oval, antero-lateral margins (**al**) formed as a curved crest armed anteriorly with a few sharp small teeth, carapace with a moderately prominent post-frontal lobe (**lb**) and post-frontal crest (**pc**), branchial regions (**br**) noticeably swollen; chelipeds (**pr1**) equal in size; second (**pr2**) to fifth (**pr5**) pereiopods stout, margins of propodus (**pro**) and dactylus (**dcy**) with prominent small spines; first gonopod (fig. B) of male strongly curved and with apical inner margin (**im**) expanded giving apex lance-shaped appearance.
COLOUR: variable, greenish to yellowish brown or almost black, limbs lighter.
CARAPACE LENGTH: maximum about 4.1 cm.
DISTRIBUTION: from France (Nice region-but now probably absent) eastward to Italy, Balkans and Greece.

The Italian Freshwater Crab inhabits freshwater streams and marginal lake waters and burrows or hides beneath stones or rocks in or adjacent to water.

GENERAL INFORMATION: the eggs of freshwater crabs hatch as miniatures of the parents, there are no larval stages as such.

At present at least eight species and twenty two subspecies of freshwater crabs of the genus *Potamon* have been described from countries bordering the Mediterranean but opinions differ among experts about the taxonomic status and nomenclature of some of these. Therefore, only three are considered here: *Potamon fluviatile fluviatile* inhabits southern Europe eastward and *P. fluviatile algeriense* North Africa. *Potamon potamios* occurs in the southern Aegean region south eastward, often some distances from water, and *P. ibericum* is found in countries bordering the Black Sea eastward.

The carapace postero-lateral margins of *P. potamios* converge more acutely than those of *P. fluviatile fluviatile* and the gonopod apex (fig. 10.24, C) is not expanded or lance-shaped; the apex of this gonopod of *P. ibericum* is cone-shaped (fig. 10.24, D).

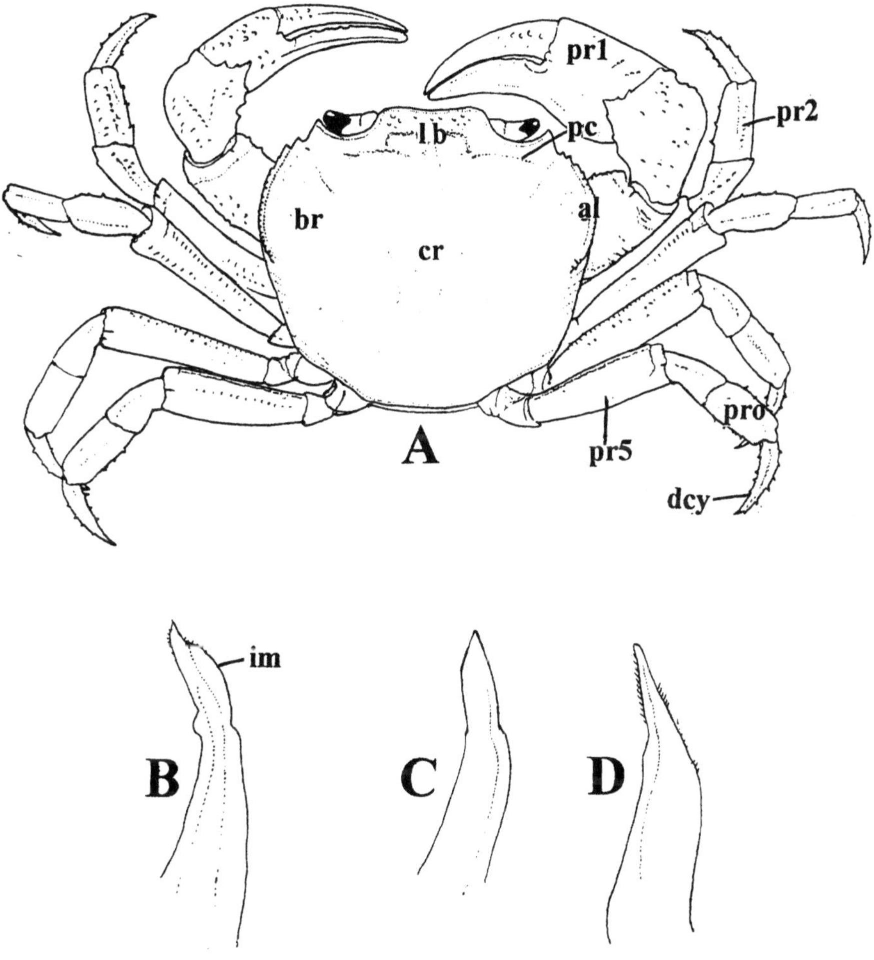

Fig. 10.24 *Potamon fluviatile fluviatile*, Italian Freshwater Crab. A. Dorsal aspect of whole crab. B. Distal part of first right gonopod of male for comparison with those of C-*P. potamios* and D-*P. ibericum*. For explanation of annotations see opposite page.

Inachus dorsettensis (Pennant, 1777)
Fig. 10.25.
COMMON SYNONYMS: none.
DERIVATION: Inachus (Gr)-a river god, son of Oceanus; dors=dorsum (Lt)-the back, set=seta (Lt), ensis (Lt) belonging to, thus alluding to the setal covered carapace surface.
VERNACULAR NAMES: Scorpion Spider Crab, Pennant's Spider Crab (UK); Stankelbenskrabbe (Sc); Gestekelde Sponspootkrabe (Nd); Crabe Fantôme (Fr); Araña de Mar (Sp).
DISTINCTIVE FEATURES: carapace (**cr** fig. A) outline sub-triangular in shape but often covered with dense encrusting growths of sponges and/or hydroids [sea firs]; rostrum (**rs**, fig. B) with U-shaped cleft; orbits (**ob**) wide, eyes conspicuous; four small spines ranging across anterior part of carapace dorsal surface; chelipeds (**pr1,** fig. A) equal in size, second (**pr2**) to fifth (**pr5**) pereiopods very long and thin, second slightly thicker than other pairs, merus (**mrs**) relatively robust; cheliped propodus (**pro**) of male swollen.
COLOUR: dorsal surface of carapace reddish brown; chelipeds (particularly carpus underside) of male pink or violet.
CARAPACE LENGTH: maximum about 3.5 cm.
DISTRIBUTION: North Eastern Atlantic-from mid-Norway southwards to Spanish Sahara (West Africa). Probably whole of Mediterranean.

The Scorpion Spider Crab occurs in onshore waters from about 4 m and downwards to at least 300 m offshore and on various substrates, chiefly sand, muddy sand and weed-covered rocky beds.
GENERAL INFORMATION: females bear eggs throughout the year in North Eastern Atlantic waters and from March until September in the Mediterranean.

Scorpion Spider Crabs are sometimes seen entangled in fishing nets on quaysides. Crabs are not infrequently on display in public aquaria where they can be seen sometimes camouflaging by using the chelipeds to pluck pieces of algae or sponges from their surroundings and placing these objects on the body and limbs.

The Scorpion Spider Crab can be distinguished from the six other species belonging to the genus *Inachus* occurring in North Eastern Atlantic/Mediterranean waters by the longer than broad carapace, the well developed spines ranging across the anterior part of the carapace, the U-shaped cleft separating each half of the rostrum and the relatively robust merus of the second pair of pereiopods.

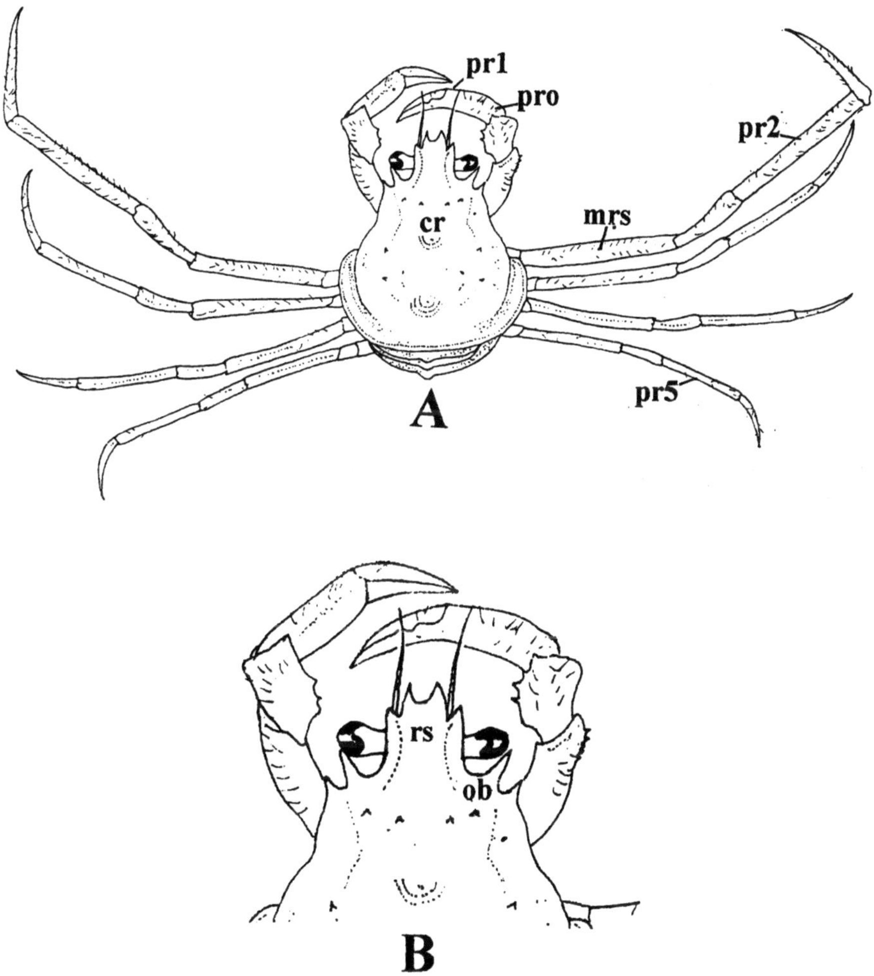

Fig. 10.25 *Inachus dorsettensis*, Scorpion Spider Crab, female. A. Whole crab from dorsal aspect. B. Anterior part of carapace showing the U-shaped rostrum and the four transversely placed spines. For explanation of annotations see opposite page.

Macropodia rostrata (Linnaeus, 1761)
Fig. 10.26.
COMMON SYNONYMS: *Stenorhynchus rostratus.*
DERIVATION: makros (Gr)-long; podos (Gr)-a foot; rostratus (Lt)-beaked.
VERNACULAR NAMES: Long-legged Spider Crab, Slender-legged Spider Crab (UK); Stankelbenskrabbe (Sc); Gewone Hooiwagenkrab (Nd); Macropode (Fr); Araña de mar patas largas (Sp).
DISTINCTIVE FEATURES: carapace (**cr,** fig. A) triangular in outline; rostrum (**rs**) prominent and slightly upturned; anterior and posterior regions of carapace dorsal surface with broad rounded elevations; eyes long and conspicuous; peduncle proximal segment (**pe**, fig. B) of antenna without spines; (chelipeds (**pr1**) stout, second (**pr2**) to fifth (**pr5**) pereiopods very long and slender, dactyls (**dcy**) of second and third very long, those of fourth and fifth shorter and curved, with very small spines on inner margin; cheliped propodus (**pro**) of male swollen.
COLOUR: dorsal surface of carapace and pereiopods grey to yellow or reddish brown.
CARAPACE LENGTH: maximum about 2.8 cm.
DISTRIBUTION: North Eastern Atlantic-from mid-Norway southwards to about Morocco (West Africa). Probably whole of Mediterranean.

The Long Legged Spider Crab occurs among algae of the lowest shore zone downwards to about 130 m on various substrates including muddy sand, algae and oyster beds.

GENERAL INFORMATION: females bearing eggs occur throughout the year in the North Eastern Atlantic and from March until September in parts of the Mediterranean.

In Mediterranean fishing ports or beaches Long legged Spider Crabs are sometimes found entangled in fishing nets laid out to dry.

This species is also often seen in public aquaria where is can be observed camouflaging itself with pieces of seaweed that it attaches to special hook-shaped hairs on carapace and limbs.

The Long Legged Spider Crab is distinguished from the seven other species of *Macropodia* inhabiting the North East Atlantic/Mediterranean region by having no spines on the proximal segment of the antennal peduncle and very small spines (spinules) on the inner margin of the fifth pereiopod dactylus.

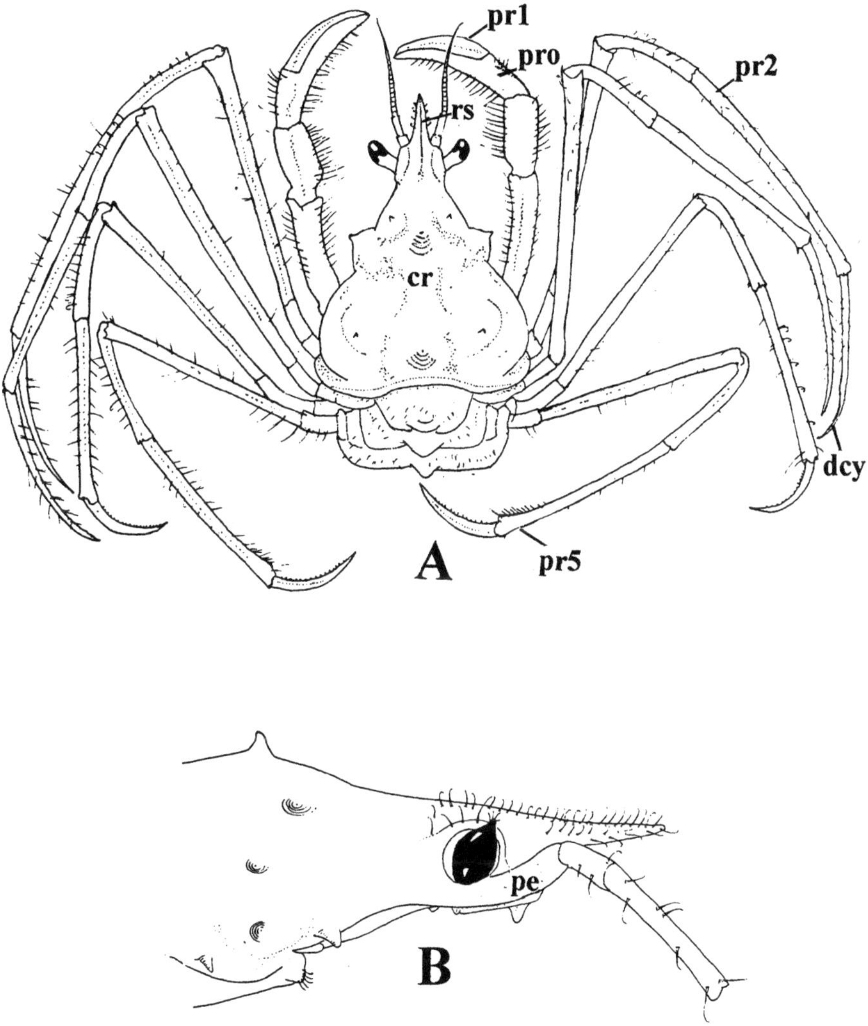

Fig. 10.26 *Macropodia rostrata*, Long Legged Spider Crab, female. A. Dorsal aspect of whole crab. B. Right lateral aspect of anterior half of carapace showing antennal peduncle situated below rostrum. For explanation of annotations see opposite page.

Maja squinado (Herbst, 1788)
Fig. 10.27, Pl.33.
COMMON SYNONYMS: none.
DERIVATION: Maia (Gr)-a large kind of crab also daughter of Atlas and mother of Mercury; squinado-a local name used in Provence for this crab.
VERNACULAR NAMES: Thorn-back Spider Crab, Common Spider Crab, Spiny Spider Crab, Corwich, King Crab (UK); Isohamahakkirapu (Fi); Maskeringskrabba, Spindelkrabba (Sc); Troldkrabbe (Dn); Grosse Seespinne (Gn); Grote Spinkrab (Nd); Araignée de Mer, Plomb, Keonid (Fr); Centolla, Boc (male), Cabra (female), Cranca, Araña (Sp); Santola (Pt); Granceola, Grancevola, Craba e'Mari (It); Gkagkala (Gr); Deniz Örümcegi (Tr).
DISTINCTIVE FEATURES: carapace (**cr**) dorsal surface very convex, outline sub-oval, antero (**al**) and part of postero-lateral margins (**ps**) with large and small spines; rostrum (**rs**) bifid and well developed and measuring about one seventh to one eight of carapace length; eyes small and often concealed by orbits (**ob**); chelipeds (**pr1**) equal in size, long in male; second (**pr2**) to fifth (**pr5**) pereiopods stout, dactyls (**dcy**) short and curved.
COLOUR: dorsal surface of carapace reddish brown, pereiopods lighter.
CARAPACE LENGTH: maximum about 22.5 cm.
DISTRIBUTION: North Eastern Atlantic-from Ireland southwards to Guinea (West Africa). Whole of Mediterranean.

The Thorn-back Spider Crab occurs in onshore waters of about 2 m depth downwards to about 150 m in the North Eastern Atlantic and between 10-140 m or deeper in the Mediterranean.

The crab usually inhabits weed-covered substrates, or sandy/shingly bottoms where weeds occur.

GENERAL INFORMATION: females bear eggs between March and October in the North Eastern Atlantic, and from April until August in the Mediterranean.

Crabs are often covered with camouflage of weed and other organisms.

UK fishermen usually regard this crab as a nuisance when numbers become entangled in nets, but some take advantage of the small but established European market for the Thorn-back Spider Crab by exporting incidental catches to Europe on a local basis.

The Thorn-back Spider Crab is distinguished from the other two species of *Maja* occurring in the southern part of the North Eastern Atlantic/Mediterranean (*M. goltziana* and *M. crispata*) by the length of the rostrum (one seventh to one eighth of the carapace length), the very convex carapace surface bearing very pointed spines, and the relatively short and straight spines on the carapace antero-lateral margins.

The vernacular names: Araignée Naine (Fr); Cabrot, Cambara, Arola (Pt); apply to the Lesser Spider Crab (UK name) *Maja crispata*, a smaller species than *M. squinado* and not specifically fished.

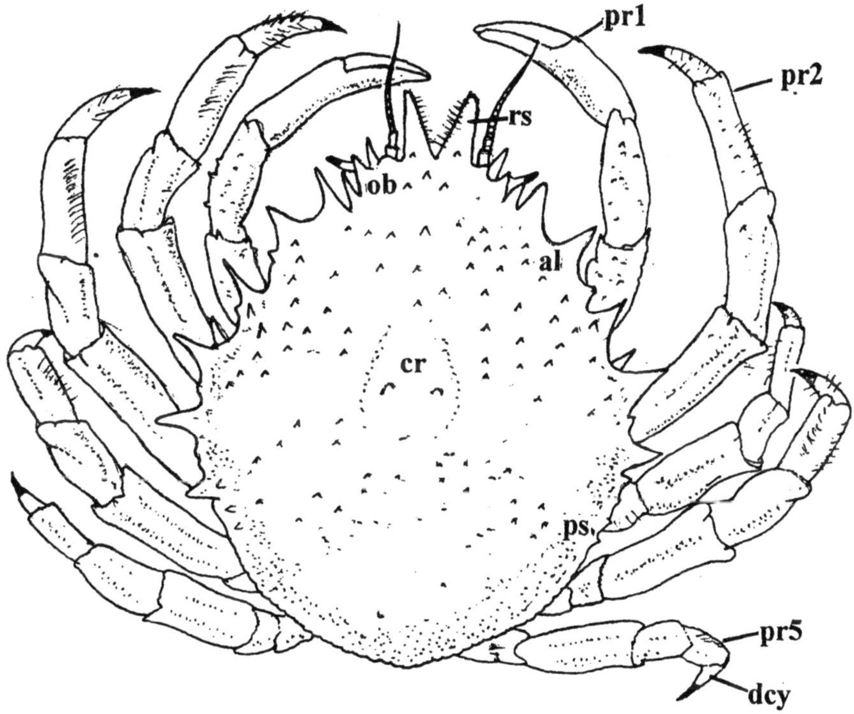

Fig. 10.27 *Maja squinado*, Thorn-back Spider Crab. For explanation of annotations see opposite page.

Pisa armata (Latreille, 1803)
Fig. 10.28.
COMMON SYNONYMS: *Pisa gibbsii.*
DERIVATION : generic name perhaps after Pisa in Tuscany Province of Italy; armatus (Lt)-armed, probably alluding to carapace spines.
VERNACULAR NAMES: Gibb's Spider Crab (UK); Gebochelde Spinkrab (Nd); Araignée à rostre pointu.(Fr).
DISTINCTIVE FEATURES: carapace (**cr**) outline sub-triangular, dorsal surface covered with dense layer of pile-like setae; antero-lateral (**al**) margins with two large spines and postero-lateral (**ps**) margins with one prominent spine, a strong spine (**sp**) on mid-posterior margin of carapace; orbital (**ob**) region with a well developed spine; rostrum (**rs**) bifid, distal portion of each half curved outwards; eyes prominent often concealed by orbits; chelipeds (**pr1**) equal in size, second (**pr2**) to fifth (**pr5**) pereiopods stout and covered with setae, dactyls (**dcy**) short and curved; cheliped propodus (**pro**) of male swollen.
COLOUR: dorsal surface of naked carapace and limbs reddish brown.
CARAPACE LENGTH: maximum about 5.5 cm.
DISTRIBUTION: North Eastern Atlantic-from about the Isle of Man region southwards to Angola (West Africa). Whole of Mediterranean.

Gibb's Spider Crab occurs in onshore waters of about 1 m depth downwards to over 160 m in the North Eastern Atlantic and to about 50 m in the Mediterranean.

The crab inhabits regions of weed covered rocky or stony substrates.
GENERAL INFORMATION: females bearing eggs occur from March until September in the North Eastern Atlantic and from July until September in the Mediterranean.

Crabs are occasionally seen entangled in nets on quayside boats in southern European and Mediterranean ports and sometimes in captivity in public aquaria.

Surfaces of carapace and limbs may have a dense covering of weed and sponges that provide an effective camouflage.

Gibb's Spider Crab is distinguished from the five other species of *Pisa* variously inhabiting the Mediterranean and North Eastern Atlantic by having on the carapace only two prominent spines behind the eyes on the antero-lateral margins and a conspicuous spine on the mid-posterior margin.

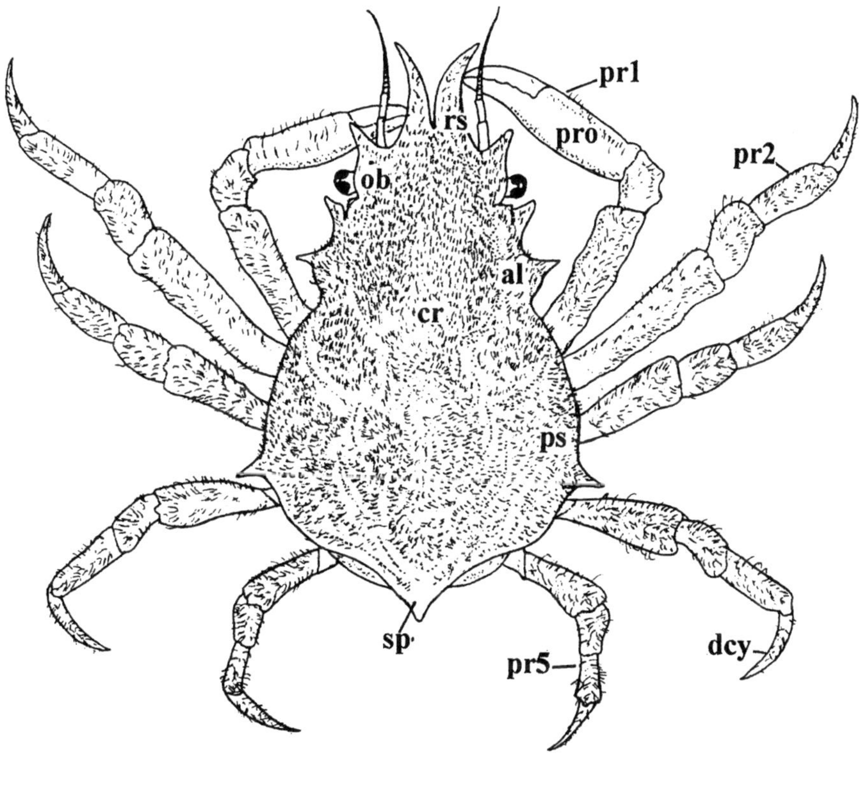

Fig. 10.28 *Pisa armata*, Gibb's Spider Crab. For explanation of annotations see opposite page.

FURTHER READING

Adema, J.P.H.M. (1991) *De Krabben Nederland en Belgie (Crustacea, Decapoda, Brachyura.* Nationaal Natuurhistorisch Museum, Leiden.

Allen, J.A. (1967) *The fauna of the Clyde Sea Area. Crustacea: Euphausiacea and Decapoda.* Scottish Marine Biological Association, Millport (pages 62-70 and 96-105 for true crabs).

Bott, R. (1970) Die Süßwasserkrabben von Europa, Asien, Australien und ihre Stammesgeschichte. *Abhandlungen der Senckenbergischen Naturforschenden Gesellschaft,* **526**, 1-338 (pages 134-139 for European species).

Clark, P.F. (1986) *North East Atlantic crabs, an atlas of distribution.* Marine Conservation Society. Hertfordshire.

Christiansen, M.E. (1969) *Marine Invertebrates of Scandinavia. No. 2. Crustacea Decapoda Brachyura.* Universitetsforlaget, Oslo.

Crothers, J. and Crothers, M. (1983) A key to the crabs and crab-like animals of British inshore waters. *Field Studies,* **5**, 753-806 (pages 765-786 and 790-806 for true crabs).

Dounas, C., Karakassis, I. and Fransen, H.J.M. (1992) Towards a revision of the East Atlantic species of the genus *Ebalia* Leach (Crustacea, Brachyura, Leucosiidae). Preliminary results. *Première Conférence Européene sur les Crustacés* (Résumé p. 48), Paris.

Ingle, R.W. (1980) *British Crabs.* British Museum (Natural History) and Oxford University Press, London and Oxford.

Ingle, R.W. (1996) *Shallow-water crabs. Synopses of the British Fauna* **25**.(Second Edition) Linnean Society of London and Estuarine and Coastal Sciences Association, Field Studies Council, Shrewsbury.

Zariquiey Alvarez, R. (1968) Crustáceòs Decápodos Ibéricos. *Investigación Pesquera,* **32**, 1-510 (pages 294-484 for hermit crabs).

11

Imported Crayfishes, Lobsters and Crabs

Freshwater crayfishes excepted, no detailed information is available recording the majority of *species* of lobsters and crabs imported into European and United Kingdom shellfish markets. Lobsters and crabs originating within the European Community are free from Custom and Excise tariffs but those imported from outside the European Union attract charges according to the species and /or tariff classifications applicable within the country of import. The United Kingdom Custom and Excise tariff and overseas trade classifications are generally non-specific for lobsters and crabs imported from non-EU countries. These are described in the Tariff as: '*Palinurus spp*, *Panulirus spp*, *Puerulus spp*, freshwater crayfishes, *Paralithodes camchaticus*, *Chionoecetes spp*' and '*Callinectes sapidus*', but the category 'other' is also listed for crustaceans not assignable to the genera listed above. A licence is required to import some species, eg. European and American Clawed Lobsters, *Homarus* (see p. 56) and for keeping in captivity both native and non-native crayfishes (see p. 179).

The vernacular names often used on custom declarations are not necessarily a totally reliable indication of the particular species imported and occasionally a consignment can comprise of more than one species. Also identifications made by exporters or importers and by those lacking taxonomic knowledge can be sometimes incorrect. Added to this is the ongoing discovery of new species, particularly in the Indo-Pacific region, from where Europe receives the majority of imported lobsters. Confusion can also arise when generic or specific names of lobsters and crabs are changed by taxonomists or when a species is known by several alternative vernacular names.

11.1 FRESHWATER CRAYFISHES

11.1.1 Review of species

At least three crayfish species have been introduced from America into Europe for aquaculture; all belong to the superfamily Astacoidea (see p. 27).

The most important of these is the Signal Crayfish *Pacifastacus (Pacifastacus) leniusculus leniusculus* (family Astacidae). This crayfish was originally imported into Europe to supplement stocks of native Noble Crayfish that had become decimated by crayfish plague. Signal Crayfishes are now cultured in Europe and the UK for local consumption or for supplying fish markets. Some have escaped from crayfish farms and feral populations are now established in European and UK rivers.

The Red Swamp or Louisiana Crayfish *Procambarus* (*Scapulicambarus*) *clarkii* (family Cambaridae) is imported into Europe from the USA. In Louisiana it is harvested in large quantities and provides an important commercial revenue. The crayfish is now cultured for food in many parts of the World. When this species was introduced into Spain and Portugal it caused damage to rice crops. It was introduced into Africa in an attempt to eradicate the water snails that provide an intermediate hosts for the parasite causing schistosomiasis. The Red Swamp Crayfish appears able to become establish in countries that have suitably warm climates, it is spreading into several regions of Europe and is being offered for sale more frequently in fish markets and in shops trading ornamental fishes.

The North American Striped Crayfish (*Orconectes* (*Faxonius*) *limosus*) (family Cambaridae) was introduced into Poland during the 1890's and subsequently into France with a view to potential commercial exploitation. It has now become established in both western and northern Central Europe and, at present, has very limited commercial value.

Australian species of crayfishes belong to the superfamily Parastacoidea and are collectively known as 'Yabbies'. Those of the genus *Cherax* (Family Parastacidae), are fished commercially and experimentally cultured. The Marron (*Cherax tenuimanus*) has been introduced into Europe for experimental aquaculture and the Melbourne Yabby (*Cherax destructor*) is occasionally imported by traders in ornamental fish. Both of these crayfishes are able to live out of water for considerable periods, demonstrating an adaptive feature to the drought conditions they often experience in the natural habitat.

11.1.2 Distinguished from native species

The Signal Crayfish is easily distinguished from all native species by the broad chelipeds (**pr1** in fig. 11.1 A), that have a conspicuous white patch on the dorsal surface near the base of the dactylus; the male of this species has the first pair of pleopods narrow and tubular (fig. 11.1, B).

Males of the Striped Crayfish have hooks on the ischium (**isc**, fig. 11.2, C) of the third pair of pereiopods. These hooks are present on the ischia (**isc**, fig. 11.3, C) of the third (**pr3**) and fourth (**pr4**) pairs of pereiopods of the Red Swamp Crayfish but are totally absent in all native species. The first pair of pleopods [gonopods] of the male are also quite different in shape from those of native species (compare figs 11.3, B with 4.5-4.6, B).

The first pair of pleopods (see **pl1**, fig. 1.1, C, D) are absent in all species of Yabbies (parastacid crayfishes) from Australia or New Zealand. These appendages are always present in native species although they may be quite small in some females.

11.1.3 Crayfishes and the law

In Britain the native White-footed Crayfish *Austropotamobius pallipes* (p.34) is listed on Schedule 5 of the Wildlife and Countryside Act of 1981. This forbids the capture and sale of this species.

Under the Prohibition of Keeping Live Fish (Crayfish) Order of 1986 a licence will be required to import and keep non-native crayfishes. Further information concerning current and proposed regulations can be obtained from the Ministry of Agriculture Fisheries and Food.

11.1.4 National Database for Crayfishes

Distribution and information pertaining to all crayfish species in the British Isles is monitored by David Holdich and David Rogers, Department of Life Science, The University of Nottingham, NG7 2RD to whom can be sent substantiated information about crayfishes.

Pacifastacus (Pacifastacus) leniusculus leniusculus (Dana, 1852)
Fig. 11.1, Pl.4.
COMMON SYNONYMS: *Astacus leniusculus; Potamobius leniusculus; Pacifastacus leniusculus.*
DERIVATION: pacif-denoting Pacific side of North America where species of this genus originate; astakos (Gr)-a kind of lobster or crayfish; lenius (Lt)-a fish mentioned by Pliny.
VERNACULAR NAMES: Signal Crayfish, Californian Crayfish, Simontorp Crayfish (UK), Écrevisse de Californie (Fr); Signalkrebse (Gr); Cangrejo de Señal.
DISTINCTIVE FEATURES: rostrum (**rs,** fig. A) with a distinctly smooth longitudinal median ridge, second post-orbital ridge (**pd**) very small; chelipeds (**pr1**) noticeably robust and broad; male-distal part of first pleopod (**pl1,** fig. B) narrow and without apical lobes.
COLOUR: dorsal surface brown to greenish brown, ventral surface often bright orange/red; basis of cheliped dactyl and propodal 'finger' with conspicuous white oval patch often edged with blue, a feature that'signals' easy recognition of this species.
TOTAL LENGTH: maximum 12-14 cm.
DISTRIBUTION: this subspecies originates from North America west of the Rocky Mountains. It was introduced into Europe during the 1960's and feral populations are now established in Sweden, Finland, France, Spain, USSR and the British Isles.

In its natural habitat the Signal Crayfish occurs in slow flowing rivers and streams but has adapted to reservoirs and lakes and will tolerates slightly saline water.
GENERAL INFORMATION: the Signal Crayfish has a life span in excess of nine years and grows faster than any of the native European species. This feature, and its resistance to crayfish plague (p. 31), has made it a particularly suitable crayfish for farming and for restocking lakes from which populations of native crayfishes have been eliminated by plague.
The species is occasionally offered for sale by ornamental fish retailers.

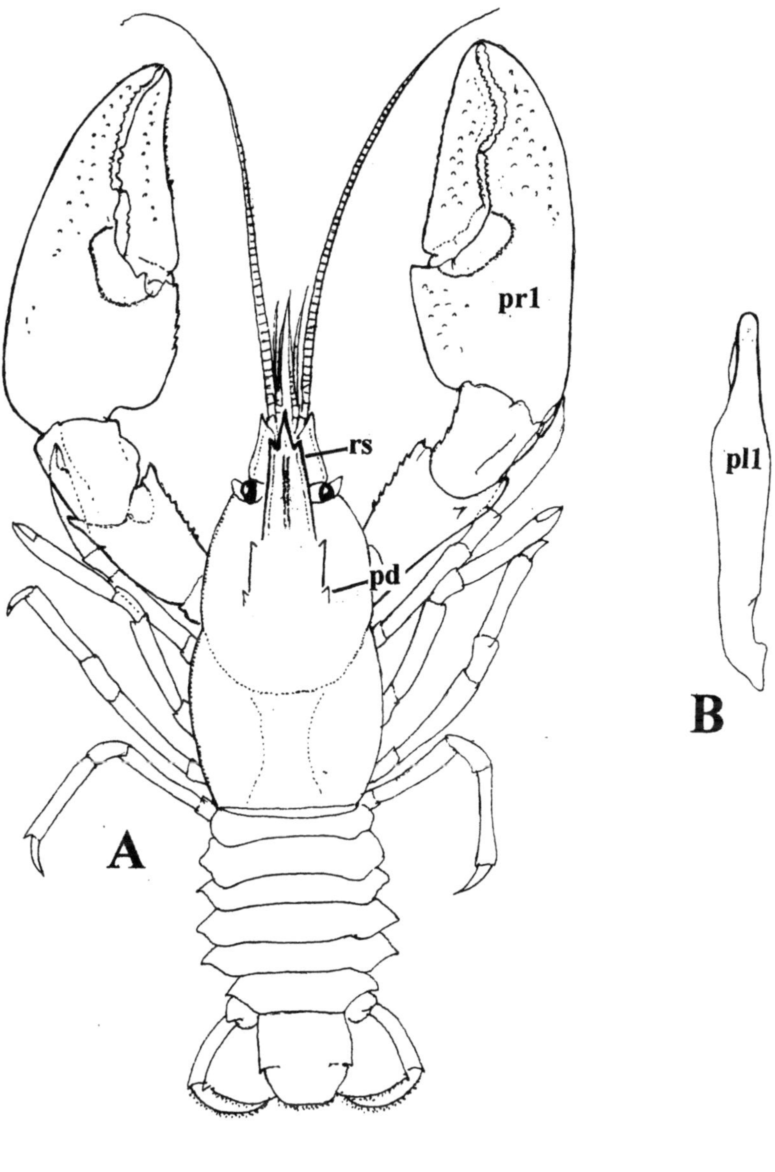

Fig. 11.1 *Pacifastacus (Pacifastacus) leniusculus leniusculus*, Signal Crayfish. A. Dorsal aspect of whole crayfish. B. First pleopod of male from right side of body and from ventral aspect. For explanation of annotations see opposite page.

Orconectes (Faxonius) limosus (Rafinesque, 1817)
Fig. 11.2, Pl. 5.
COMMON SYNONYMS: *Cambarus affinis*; *Orconectes limosus.*
DERIVATION: orkos (Gr)-segment; nekter (Gr)-a swimmer; limosus (Lt)-mud or slime (denoting preference for muddy substrates).
VERNACULAR NAMES: Striped Crayfish, American River Crayfish, Camber Crayfish (UK); Amerikanischer Flusskrebs (Gn); Cangrejo de Rio Americano (Sp).
DISTINCTIVE FEATURES: margins of rostrum (**rs,** fig. A) almost parallel, only a single pair of post-orbital ridges (**pd**) present; antero-lateral region (**al**) of carapace with one prominent spine and usually 2-3 smaller spines; inner margin of cheliped carpus (**crp**) with a prominent curved spine; male-distal lobes of first pleopod (**pl1,** fig. B) diverging; ischium (**isc,** fig. C) of third pereiopod with a distinct hook, first (**pl**) and second pleopods never concealed by hairy mass arising from sides of sternum.
COLOUR; olive green, abdomen sometimes with transverse bands of reddish brown.
TOTAL LENGTH: average 9 cm, rarely reaching 11 cm.
DISTRIBUTION: A native of Eastern USA (Maine to Virginia). Introduced into Europe in 1890's to restock tributary of River Oder, Poland and to replace depleted stocks of Noble Crayfish. Successfully introduced into France between 1911-1913 and now established in western and northern Central Europe where it occurs from North Eastern Poland to Lithuanian border and westward to South Western France, in Germany it extends as far south as the Bavarian border.
Habitats of the Striped Crayfish range from small streams to shallow regions of lakes and ponds. The species occurs on gravel to muddy substrates into which it readily burrows.
GENERAL INFORMATION: A productive species but with a short life span. Tolerates greater degrees of pollution than native European species of crayfishes and also believed to have a marked resistance to crayfish plague. The Striped Crayfish is occasionally offered for sale by ornamental fish retailers.

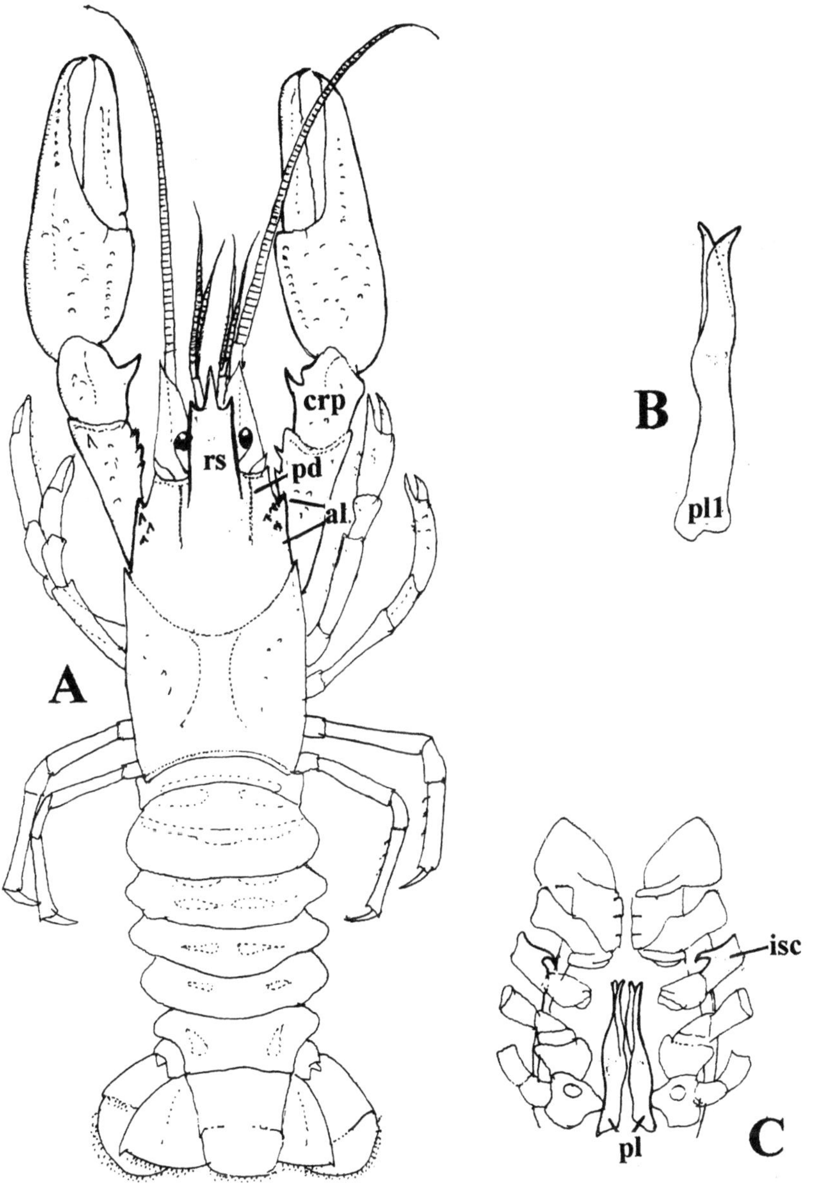

Fig. 11.2 *Orconectes (Faxonius) limosus*, Striped Crayfish. A. Dorsal aspect of whole crayfish. B. First pleopod of male from right side of body and from ventral aspect. C. Sternum of male from ventral aspect. For explanation of annotations see opposite page.

Procambarus (Scapulicambarus) clarkii (Girard, 1852)
Fig. 11.3, Pl. 6.
COMMON SYNONYMS : *Cambarus clarkii; Procambarus clarkii.*
DERIVATION: pro (Gr)-before; cambarus probably derived from kammaros (Gr)-a sea crab; clarkii species dedicated to H.J Clark the collector employed by the United States Boundary Commission.
VERNACULAR NAMES: Red Swamp Crayfish, Louisiana Swamp Crayfish, Red Crayfish (UK), Écrevisse Rouge de Marais (Fr).
DISTINCTIVE FEATURES: margins of rostrum (**rs,** fig. A) strongly converging distally, a single pair of post-orbital ridges (**pd**) present; antero-lateral (**al**) regions of carapace without prominent spine, at most with small spinules; inner margin of cheliped carpus (**crp**) with conspicuous curved spine; male-distal parts of first pleopods (**pl1,** fig. B) with angular shoulder; ischium (**isc,** fig. C) of third (**pr3**) and fourth (**pr4**) pereiopods with distal hooks, pleopods never concealed by hairy mass arising from sides of sternum, basal part of first pleopods (**pl**, fig. C) asymmetrical.
COLOUR: dark crimson, tubercles, spines and distal parts of chelipeds bright red, second to fifth pereiopods orange; abdominal segments often with thin transverse and short vertical bluish bands.
TOTAL LENGTH: maximum about 12 cm.
DISTRIBUTION: a native of America occurring from southern Illinois to northern Mexico. Introduced into other parts of USA and into western Mexico, Costa Rica, Dominica, Brazil, Berlize and Ecuador. Imported into UK, Spain, France, Portugal, Cyprus, Kenya, Uganda, China, Taiwan, Japan and Hawaii.

The natural habitat of the Red Swamp Crayfish is fresh to brackish water swamps and marshes subjected to seasonal flooding. It is often now found in irrigated regions, eg. rice fields, and has become also adapted to living in streams and lakes. It readily burrows to depths of 40-90 cm down to the water table during dry periods.
GENERAL INFORMATION: this crayfish was used by Houmas Indians as their war totem.

Probably eaten by early Louisianan immigrants (about 1700) and continuously fished their since about 1900. Commercially cultured since about 1950 in USA and now by many other countries.
Now frequently offered for sale by ornamental fish retailers.

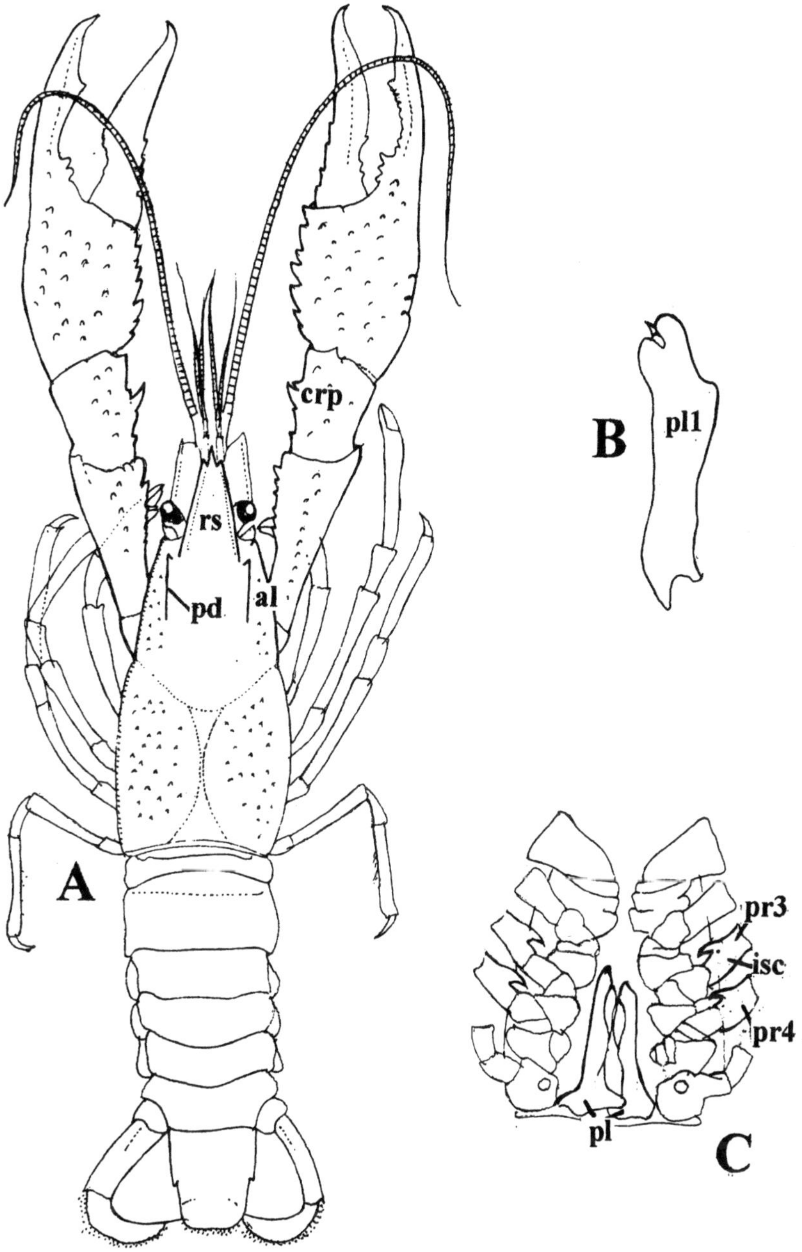

Fig. 11.3 *Procambarus (Scapulicambarus) clarkii*, Red Swamp Crayfish. A. Dorsal aspect of whole crayfish. B. First pleopod of male from right side of body and from ventral aspect. C. Sternum of male from ventral aspect. For explanation of annotations see opposite page.

Cherax destructor Clark, 1936
Fig. 11.4, Pl. 7.
COMMON SYNONYMS: *Parachaeraps bicarinatus*; *Cherax albidus*; (see Sokol, 1988 in Further Reading section for discussion on current confused taxonomy).
DERIVATION: chaer (Gr)-fish of the Nile; destructor-alluding to its ability to destroy river banks by making numerous burrows.
VERNACULAR NAMES: Melbourne Yabby.
DISTINCTIVE FEATURES: first pair of pleopods absent in both sexes; keel on sternum not strongly raised; margins of rostrum (**rs**) slightly curved and without spines; post-orbital ridges (**pd**) only slightly raised; carapace areolae (**ar**) converging in middle part; punctations on carapace prominent; carpus (**crp**) of chelipeds with blunt spines; telson (**tl**) without medial spines; basis (**bas**) of uropods without a process, at most with a minute spine; at least endopods (**en**) of uropods distally membranous.
COLOUR: blue with greenish tinges on mid-dorsal part of carapace, undersides of chelipeds white.
DISTRIBUTION: the Melbourne Yabby is widespread in Victoria and extends northwards into inland New South Wales and Queensland, Australia.

This crayfish burrows, during drought conditions, downwards to the level of the water table.

TOTAL LENGTH: average 16 cm.
GENERAL INFORMATION: the Melbourne Yabby is fished by amateur fishermen in Australia. It is of interest to aquaculturists along with the Nurrabeil Yabby, *Cherax albidus* and the large Marron, *Cherax tenuimanus*. The crayfish is occasionally imported into the UK for the aquarium trade.

The newly hatched young of this species and other parastacid crayfishes attached themselves to the female's pleopod setae by the curved spines on the dactyls of the fourth and fifth pereiopods in contrast to crayfishes of the Northern Hemisphere that use the chelipeds.

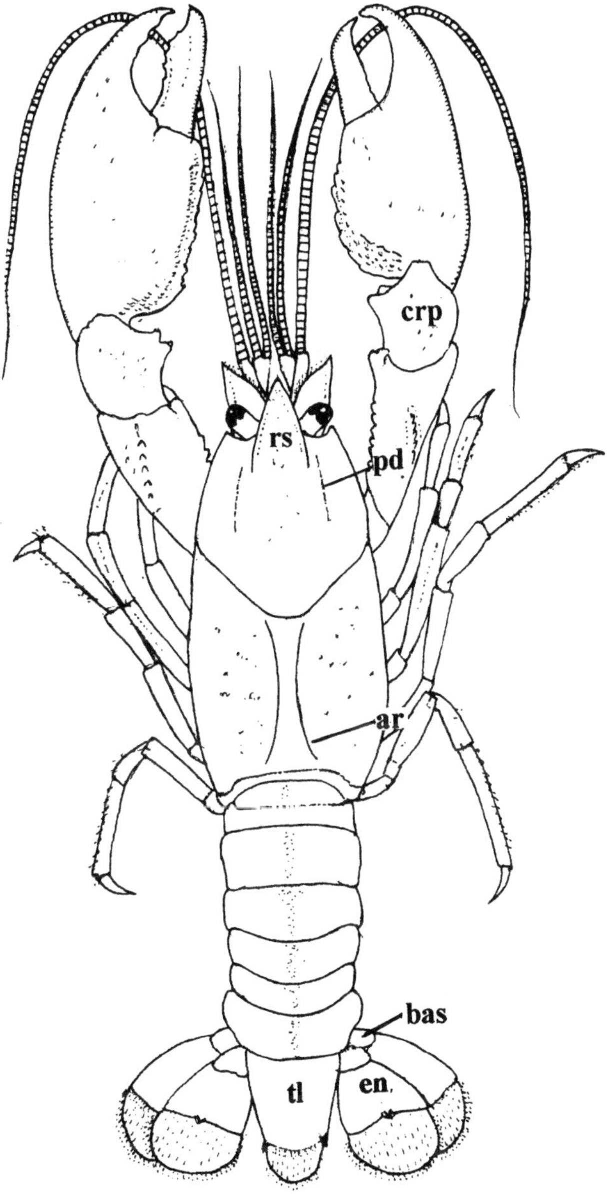

Fig. 11.4 *Cherax destructor*, Melbourne Yabby. For explanation of annotations see opposite page.

Cherax tenuimanus (Smith, 1912)
Fig. 11.5, Pl. 8.
COMMON SYNONYMS: none.
DERIVATION: FROM charax (Gr)-a pointed stake ie. a thing that scratches, perhaps alluding to sharp rostrum; tenui (Lt)-slender; manus (Lt)-a hand.
VERNACULAR NAMES: Marron.
DISTINCTIVE FEATURES: first pair of pleopods absent in both sexes; keel on sternum not strongly raised; margins of rostrum (**rs**) straight, with spines, and extending for some distance backwards onto carapace; post-orbital ridges (**pd**) prominent; carapace with medial ridge; carapace areolae (**ar**) converging in middle part; punctations on carapace present; carpus (**crp**) of chelipeds with a sharp spine; telson (**tl**) with a pair of medial spines; basis of uropods (**bas**) without a process, sometimes with a small spine, endopods (**en**) distally membranous.
COLOUR: brownish black with patches of red, chelae dorsal surface with wide red band.
DISTRIBUTION: south western region of Western Australia.

The Marron prefers to live in permanent water bodies and burrows downwards to the water table in drought conditions.

TOTAL LENGTH: maximum 38 cm.
GENERAL INFORMATION: the Marron is the third largest freshwater crayfish known and is exceeded in length by the the Tasmanian Crayfish, *Astacopsis gouldi* and the the Murray Cray, *Euastacus armatus* both reaching at least 45 cm in total length.

This crayfish is fished by amateur fishermen in Australia along with other species.

Aquaculture potential of this species has been under investigation in Australia for a number of years. The crayfish is exported to various countries including those of Europe and UK for experimental culture.

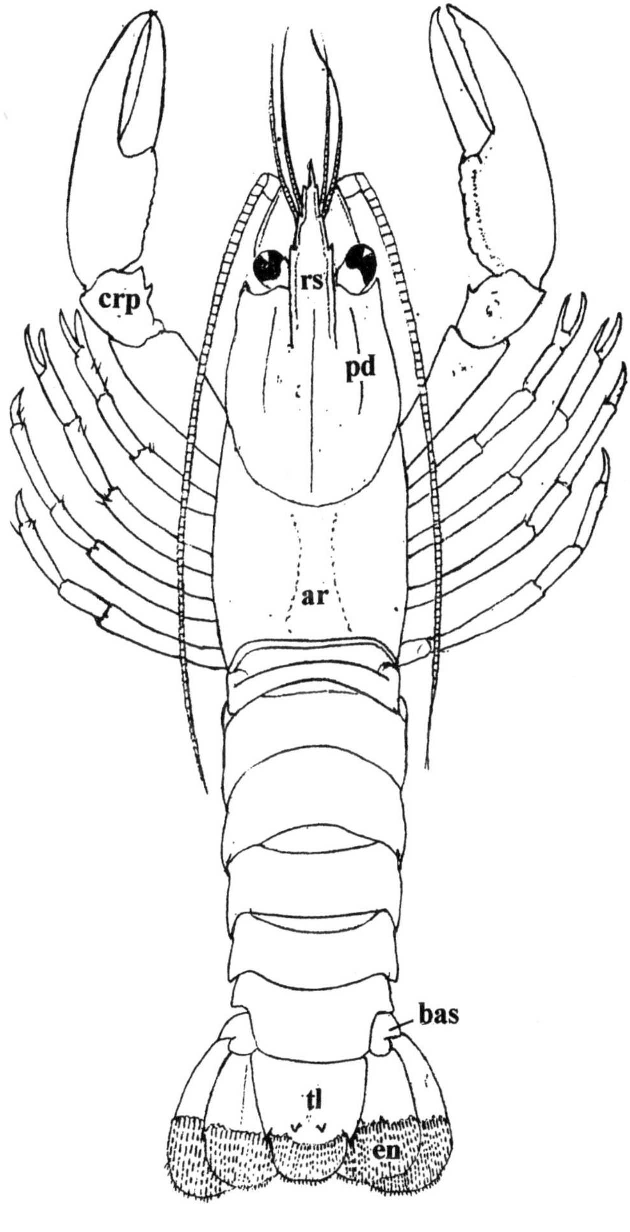

Fig. 11.5 *Cherax tenuimanus*, Marron. For explanation of annotations see opposite page.

11.2 CLAWED LOBSTERS

Two types of clawed lobsters are imported into UK and European markets: (1) homarid clawed lobsters, represented by one species, the American Clawed Lobster *Homarus americanus*; (2) nephropid clawed lobsters belonging to the genus *Metanephrops.*

11.2.1 American Clawed Lobster (*Homarus americanus*)

(a) Fisheries and marketing

This lobster is extensively fished off the coasts of Newfoundland southwards to North Carolina, chiefly with creels set inshore but it is also trawled in offshore waters. Lobsters are exported live to Europe where they are often maintained in holding tanks by importers before being sold on to customers. Brine-frozen 'bubble-packed' whole or parts of lobsters and Canadian canned lobster meat are alternative products but only limited quantities appear to be imported.

American clawed lobsters occasionally occur in British waters. These have probably escaped from commercial holdings of live lobster. The importation of American Clawed Lobsters into Europe are subjected to legislation to reduce the risk of introducing the highly virulent bacterial disease Gaffkaemia (of which this lobster can be a latent carrier) into native lobster populations. Regulations vary from country to country. In the UK controls were introduced in 1981 and licences are required from the Ministry of Agriculture Fisheries and Food for importing, moving and storing live or processed lobsters if they are to be deposited in designated coastal areas of England, Wales and Scotland.

(b) Distinguished from native European Clawed Lobster

The following combined features distinguished the two species of lobsters. The American Clawed Lobster (*Homarus americanus*) usually has a prominent distal spine (**sp**, fig. 11.6, A) on the lower margin of the rostrum. The chelae (fig.11.6 B & Pl. 10) are very broad and the spines blunt. The undersides of the chelipeds, the other limbs, and often the cephalothorax is deep orange to red. Vernacular names: American Lobster (UK); Homard American (Fr); Bogavante americano (Sp).

The European Clawed Lobster (*Homarus gammarus*). is without a distal spine on the rostral lower margin (fig. 11.6, C). The chelae (fig. D) are not very broad and the spines sub-acute. The undersides of the chelipeds, the other limbs, and the cephalothorax varies from ivory to yellowish green.

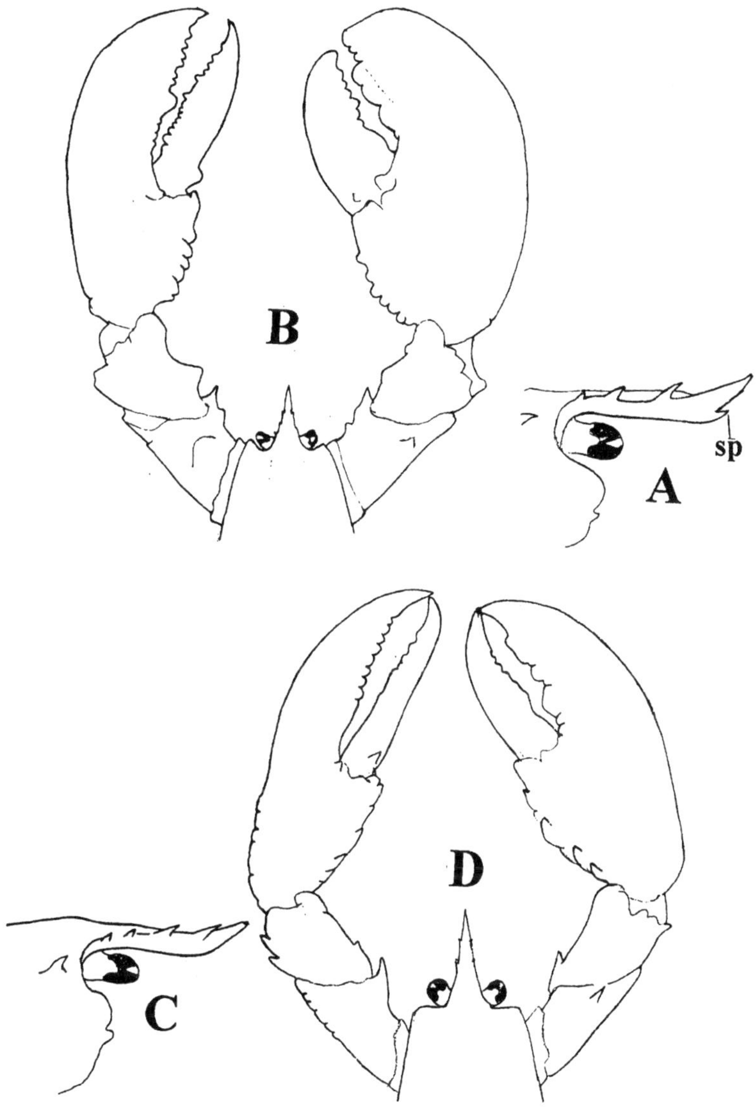

Fig. 11.6 A, C. Rostrum from right lateral aspect. B, D. Anterior part of carapace and chelipeds. A, B. American Clawed Lobster, *Homarus americanus*. C, D. European Clawed Lobster, *Homarus gammarus*.

11.2.2 Nephropid clawed lobsters

(a) Metanephropid clawed lobsters

A group of nephropid clawed lobsters belonging to the genus *Metanephrops* and comprising seventeen species occur in tropical western Atlantic and the Indo-Pacific regions. Increasing commercial interest has resulted in at least one species, the Andaman Clawed Lobster (*Metanephrops andamanicus,* being offered for sale in European and UK markets and other species may be eventually marketed here. All are small, rarely exceeding about 20 cm in total length and are similar in appearance to the Norway Clawed Lobster (*Nephrops norvegicus*) extensively fished in North Atlantic waters.

(b) Distinguished from native Norway Clawed Lobster

The following comparative features will enable the reader to readily distinguish our native Norway Clawed Lobster from imported species of metanephropid clawed lobsters.

Metanephropid clawed lobsters (Fig. 11.7, A)
Spines on rostral *margins* (**rs**) extending backwards onto anterior half of carapace. Antennal scale (**as**) short and broad. Antennal spine (**ae**) continued backwards as a conspicuously developed ridge. Chelipeds (**pr1**) narrow, almost identical in shape and size; any small bosses present on distal propodal (**pro**) and dactyl (**dcy**) inner margins present on both chelipeds.

Norway Clawed Lobster (Fig. 11.7, B)
Spines on rostral *margins* (**rs**) not extending backwards onto anterior half of carapace; replaced in that region by a pair of noticeably diverging spinulate low ridges (**rg**). Antennal scale (**as**) long and narrow. Antennal spine (**ae**) not continued backwards as a ridge. Chelipeds (**pr1**) broad, slightly unequal in size; one (crushing claw) with well developed bosses on distal propodal (**pro**) and dactyl (**dcy**) inner margins.

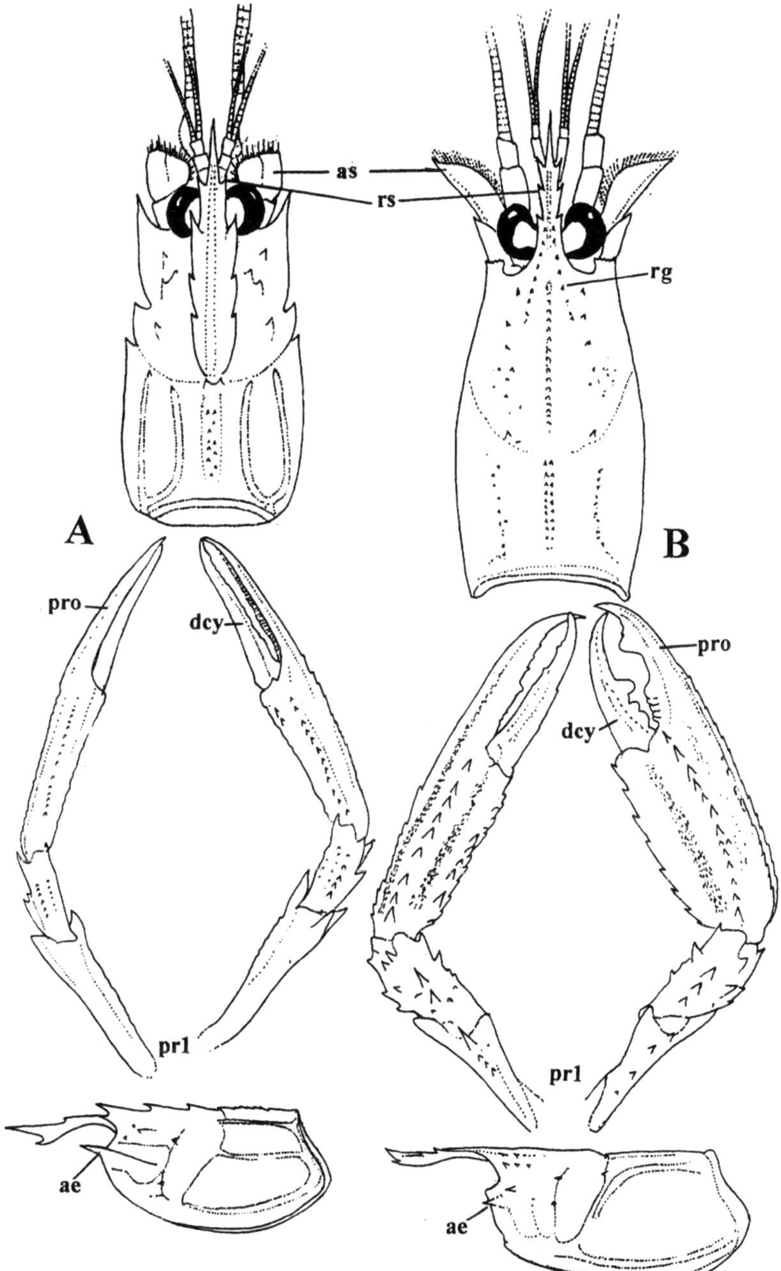

Fig. 11.7 A. *Metanephrops andamanicus*, Andaman Clawed Lobster. B. *Nephrops norvegicus*, Norway Clawed Lobster. For explanation of annotations see opposite page.

(c) Currently exploited Indo-Pacific metanephropid clawed lobsters

Sometimes termed Lobsterettes, at least eight species of metanephropid clawed lobsters appear to be fished on a regular commercial basis although it is not certain whether or not significant imports of any except the Andaman Clawed Lobster (*M. andamanicus*, Pl. 12) have reached European markets.

M. armatus (Armoured Clawed Lobster) is sold in Taiwan markets, *M. formosanus* (Formosan Clawed Lobster) in Chinese fish markets, and *M. japonicus* (Japanese Clawed Lobster) and possibly *M. sagamiensis* (Sculptured Clawed Lobster) are marketed fresh or frozen in Japan. *M. mozambicus* (African Clawed Lobster) is commercially trawled off South Africa and *M. velutinus* (Velvet Clawed Lobster) is marketed in Australia. It is possible that the last two mentioned species may have been imported into Europe but incorrectly identified as the Andaman Lobster. The seventh species, *M. thomsoni* (Red-banded Clawed Lobster) is sold in Korean and Taiwan markets. Features for distinguishing all these species are compared in table 11.1.

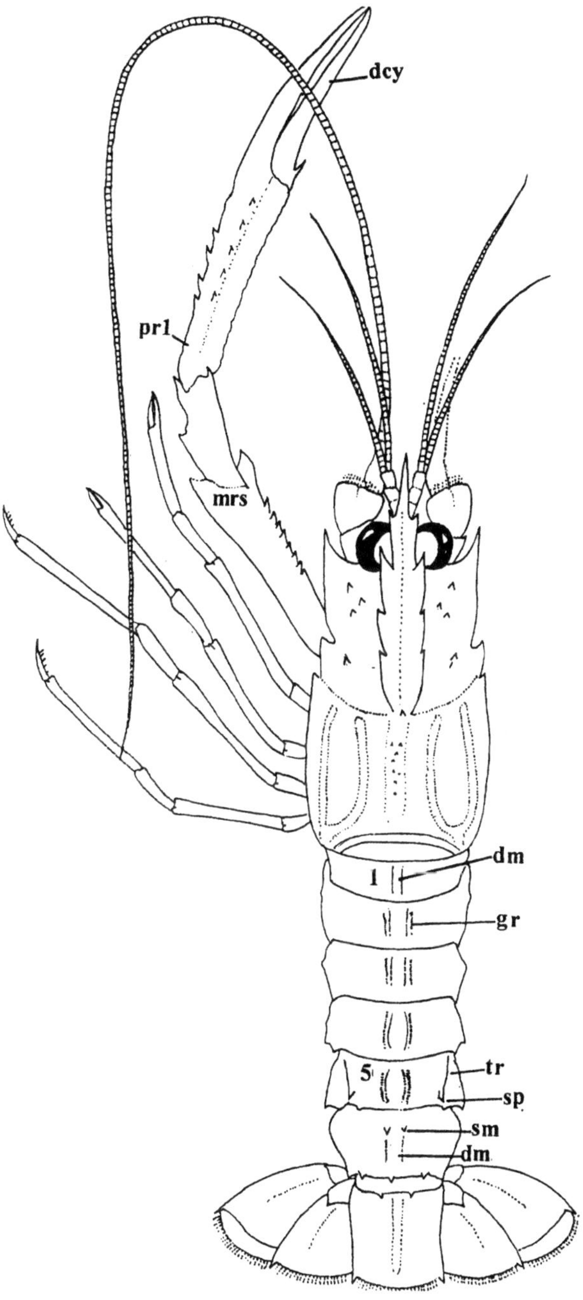

Fig. 11.8 Features used to identify the potentially imported species of metanephropid clawed lobsters listed in table 1.1.

Table 11.1 Comparative features of metanephropid clawed lobsters

Species	*M.andamanicus*	*M.armatus*	*M.formosanus*	*M.japonicus*
Feature fig. 11.8				
1st abdominal segment (1), dorsomedial ridge (dm)	absent	absent	absent	present
2nd-5th abdominal segments, grooves (gr) on each side of dorsomedial (dm) ridge	conspicuous	absent	absent	absent
5th abdominal segment (5), terminal spine (sp) on tergite ridge (tr)	not distinct	present	absent	present
6th abdominal segment, submedial spines (sm) on dorsomedial ridge (dm)	absent	present	absent	present
Chelipeds (pr1)	ridged and spinosed	ridged and spinosed	ridged and spinosed	ridged and spinosed
Spine on outer base of cheliped dactylus (dcy)	absent	present	present	present
Spines on inner margin of cheliped merus (mrs)	small	small	absent	conspicuous

M.sagamiensis	*M.mozambicus*	*M.velutinus*	*M.thomsoni*
absent	absent	absent	absent
absent	absent	absent	absent
absent	absent	indistinct	present but blunt
absent	absent	absent	absent
ridged and spinosed	ridged and spinosed	ridged and spinosed	with granular surface
absent	absent	absent	absent but tubercles present
very conspicuous	small	absent	absent

11.3 CRAWFISHES

11.3.1 Confusion about vernacular names

It is probable that more confusion occurs over vernacular names applied to this group of lobsters (all of which belong to the family Palinuridae) than any other edible crustaceans. Crawfishes are often marketed under alternative names such as 'Spiny Lobsters, Crayfishes, Marine Crayfishes' or 'Rock Lobsters' and many species are known by more than one vernacular name in most European languages, also local names of country of origin are sometimes cited.

Those involved with shellfish who wish to be reasonably specific about their importations would no doubt agree to a restricted use of vernacular names at both generic and specific levels. It is suggested here that the term 'crawfish' should be used only for species belonging to the family Palinuridae to unambiguously distinguish them from freshwater crayfishes and clawed lobsters. For the crawfishes described in the following pages the term 'Rock Crawfish' is restricted to species belonging to the genus *Jasus*, 'Spiny Crawfish' to species of *Panulirus* and *Palinurus*, (the generic names *Panulirus* and *Palinurus* must not be confused, they are quite distinct genera) and 'Whip Crawfish' to the species of *Puerulus*.

11.3.2 Fisheries for imported crawfishes

Although in terms of World catch crawfishes rate second to those of temperate water clawed lobsters they nevertheless, support important and highly profitable local fisheries. The majority of species are fished in sub-tropical and tropical waters of Atlantic and Indo-Pacific regions.

Generally the average catches of crawfishes are smaller than those of clawed lobsters, but the considerable size of many species (eg. 60 cm for the Green Rock Crawfish, (figs 11.15 & 11.16, E) and their often spectacular colour makes them very saleable items when displayed in restaurants or in local fish markets. In the Far East a few of the more colourful species are offered to tourists.as dried and mounted specimens

Throughout the World probably in excess of thirty species of crawfishes are caught in sufficient quantities to be traded. It is chiefly only frozen 'tails' (abdomens) that are imported as this part contains the consumable meat. Unlike clawed lobsters there is no meat of any significant value in the chelipeds. For marketing purposes crawfishes tails are often superficially classified into 'cold water' and 'warm water' types, 'cold water' tails commanding the higher prices. With few exceptions 'cold water' tails are from species of Rock Crawfishes (*Jasus*) and 'warm water' tails from Spiny Crawfish (*Panulirus*) species.

At least nineteen species of spiny crawfishes belong to the genus *Panulirus* and all grow to a sufficient size to provide marketable meat. Diverse fishing methods are used to catch these crawfishes. Along coastal regions of South East Asia they are fished by local communities and collected by diving and hooking them out from crevices in rocks, or by trapping in tangle nets or in lobster pots. These catches are usually sold to local hotels, restaurants and exporters. Because the majority of spiny crawfishes appear to inhabit rocky substrates trawling is usually least effective or sometimes impracticable and those that reach the European markets in quantity are often the produce of larger commercial organisations that use chiefly pots or creels for obtaining catches.

Similar to clawed lobsters there is very little information available detailing the species of crawfishes that are currently imported into UK and European markets but invariably they will be represented by the genera *Panulirus* (spiny crawfishes), *Jasus* (rock crawfishes) and *Puerulus* (whip crawfishes).

11.3.3 Distinguishing genera of imported Crawfishes

Combined features for distinguishing between spiny, rock and whip crawfishes are listed in table 11.2. Although species of the spiny crawfish genus *Palinurus* are not imported, its features are included that will enable readers to distinguish the European *Palinurus* species described on pp 66-67, from imported species of *Panulirus*.

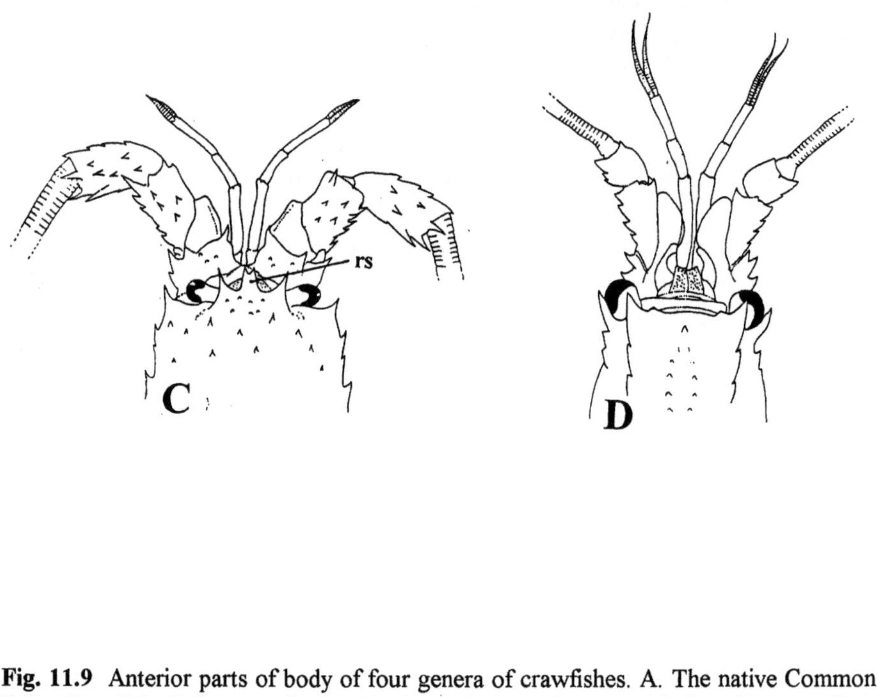

Fig. 11.9 Anterior parts of body of four genera of crawfishes. A. The native Common Spiny Crawfish, *Palinurus elephas*. B. A species of spiny crawfish *Panulirus* sp. C. A species of rock crawfish, *Jasus* sp. D. Arabian Whip Crawfish, *Puerulus sewelli*. For explanation of annotations see table 11.2.

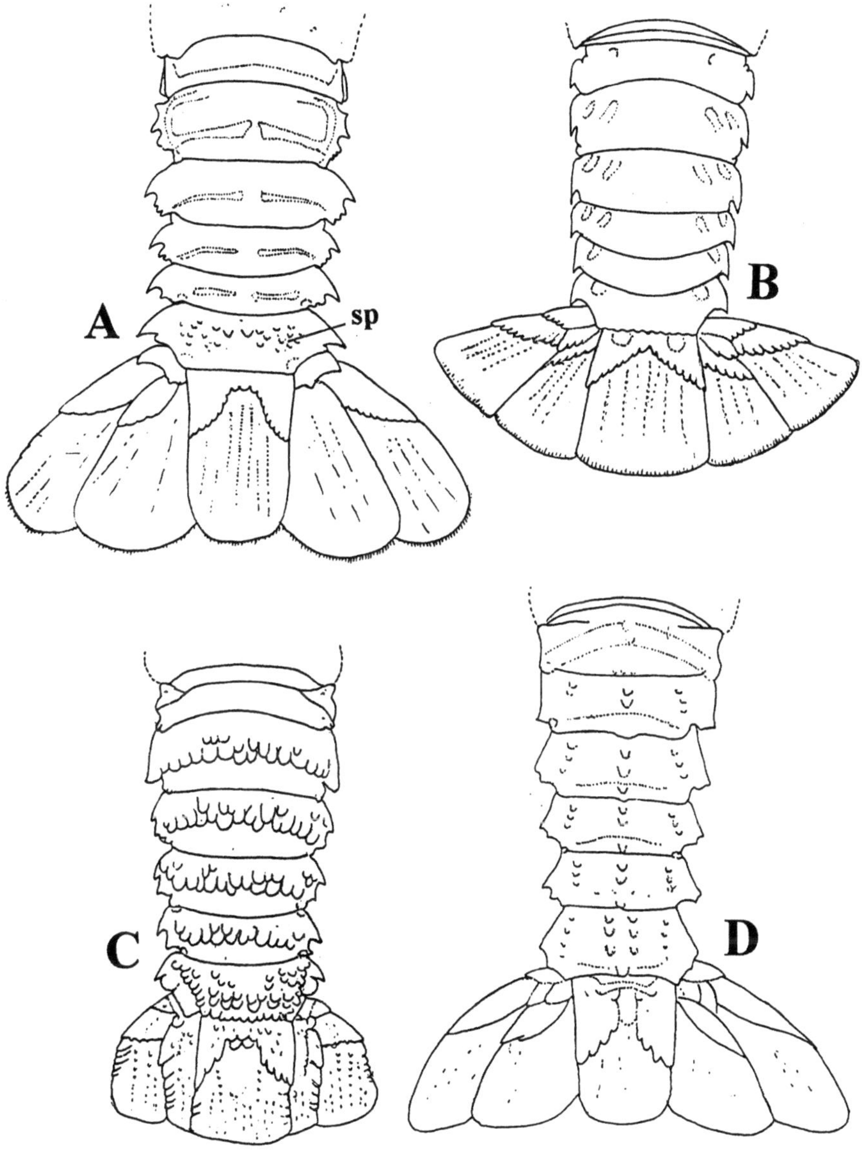

Fig. 11.10 Abdomens of four genera of crawfishes. A. Common Spiny Crawfish, *Palinurus elephas*. B. A species of spiny crawfish, *Panulirus* sp. C. A species of rock crawfish, *Jasus* sp. D. Arabian Whip Crawfish, *Puerulus sewelli*. For explanation of annotation see table 2.

Table 11.2 Distinguishing features of four genera of crawfishes.

GENUS	*Palinurus*	*Panulirus*	*Jasus*	*Puerulus*
Flagellum (fl) length (Fig.11.9, A, B) of first antenna measured against length of peduncle (pe)	shorter	longer	shorter	shorter
Antennular plate (ap, fig. 11.9, A and shaded in figs B-D)	distinctly developed but without spines on dorsal surface (fig. 11.9, A)	very obvious and with spines on dorsal surface (fig. 11.9, B)	hardly developed and without spines (fig. 11.9, C)	distinctly developed and with spines on dorsal surface (fig. 11.9, D)
Scale-like sculpturing on dorsal surfaces of abdominal segments (fig. 11.10, C)	absent	absent, but very poorly developed in *P. homarus*	present and conspicuous (fig. 11.10, C) (except *J. (S) verreauxi*)	absent
Tubercles and small spines (sp, fig. 11.10, A) on dorsal surface of sixth abdominal segment	present (fig. 11.10, A)	absent	absent (scale-like sculpturing present)	paired longitudinal row of small tubercles present (fig.11.10, D)

11.3.4 Spiny crawfishes (Genus *Panulirus*)

Imported species These spiny crawfishes will belong to the genus *Panulirus* since no species of *Palinurus* are imported. It is difficult to determine how many species of *Panulirus* are being traded in European and UK markets, or even those brought back as trophies by tourists returning from the Far East. In the author's opinion at least six species may be represented among importations. *Panulirus argus* (Caribbean Spiny Crawfish), *P. cygnus* (Australian Spiny Crawfish), *P. ornatus* (Ornate Spiny Crawfish), *P. polyphagus* (Mud Spiny Crawfish), *P. homarus* (Scalloped Spiny Crawfish) and *P. versicolor* (Painted Spiny Crawfish). The features of these species are compared in table 11.3 and illustrated in figs 11.11-11.14.

Distinguished from native species The North Eastern Atlantic Royal Spiny Crawfish, *Panulirus regius* (p. 68), occasionally seen in European markets and restaurants, has white transverse bands on the abdominal segments and therefore could be confused with two possible imported species, ie. the Mud Spiny Crawfish (*Panulirus polyphagus*) and the Painted Spiny Crawfish (*Panulirus versicolor*). However, the Royal Spiny Crawfish can be recognised by having the combined features of *four* large spines on the antennal plate and *smooth* transverse grooves, interrupted medially, on the abdominal segments.

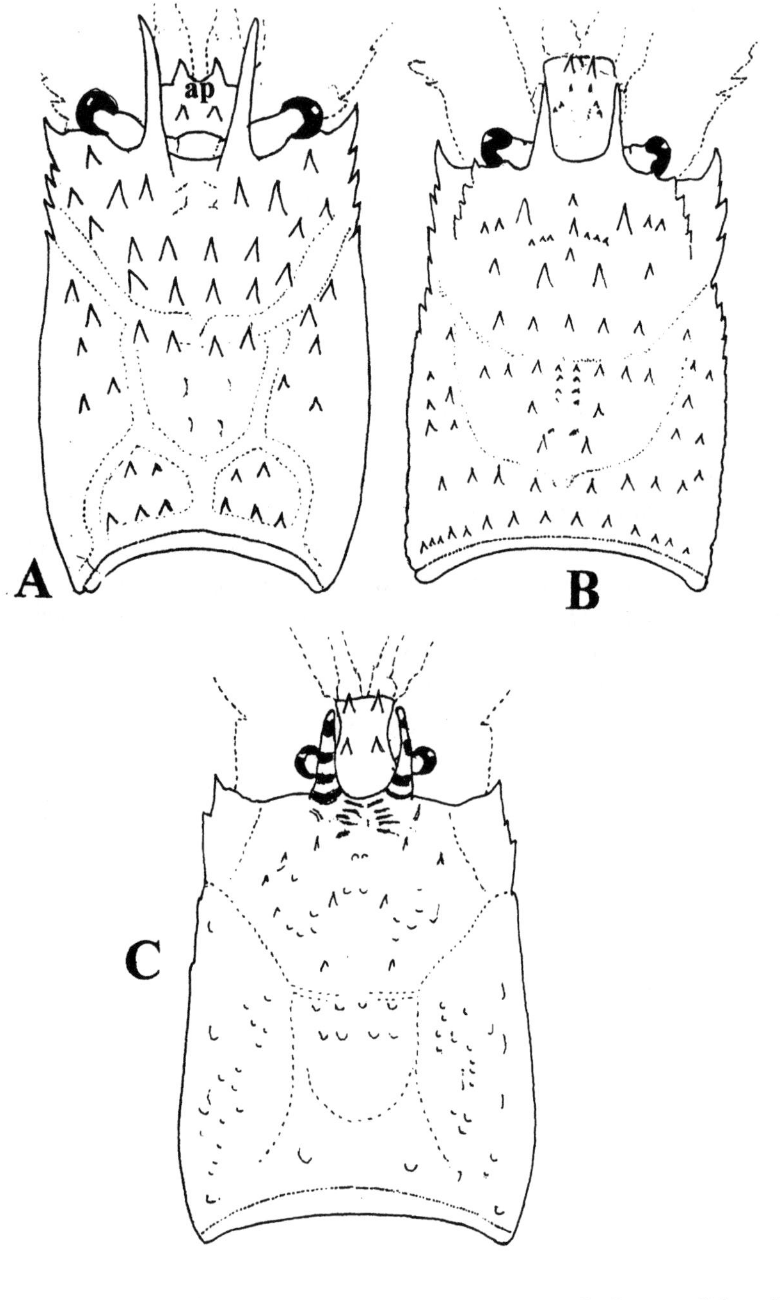

Fig. 11.11 Dorsal aspects of carapace and antennular plates (**ap**) of spiny crawfishes. A. *Panulirus argus*, Caribbean Spiny Crawfish. B. *Panulirus cygnus*, Australian Spiny Crawfish. C. *Panulirus ornatus*, Ornate Spiny Crawfish. See table 11.3.

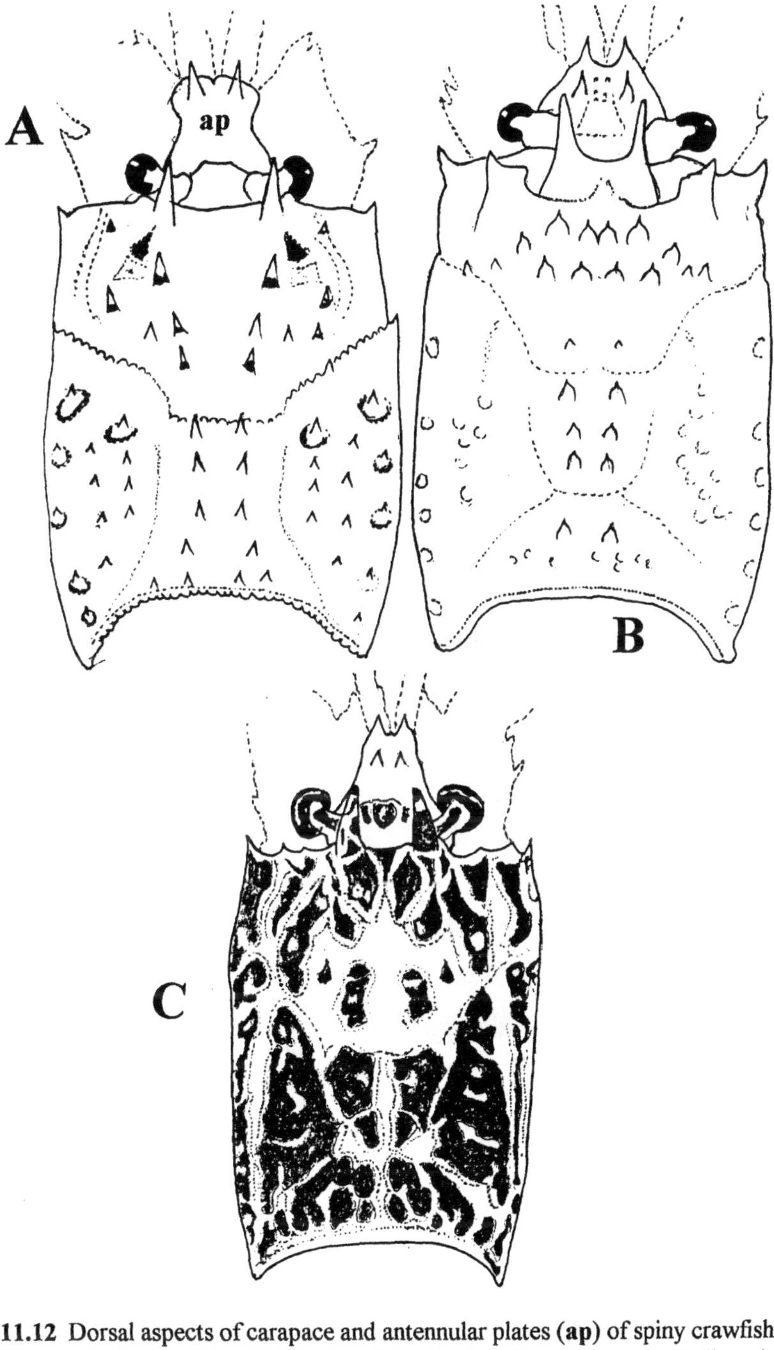

Fig. 11.12 Dorsal aspects of carapace and antennular plates (**ap**) of spiny crawfishes. A. *Panulirus polyphagus,* Mud Spiny Crawfish. B. *Panulirus homarus,* Scalloped Spiny Crawfish. C. *Panulirus versicolor,* Painted Spiny Crawfish. See table 11.3.

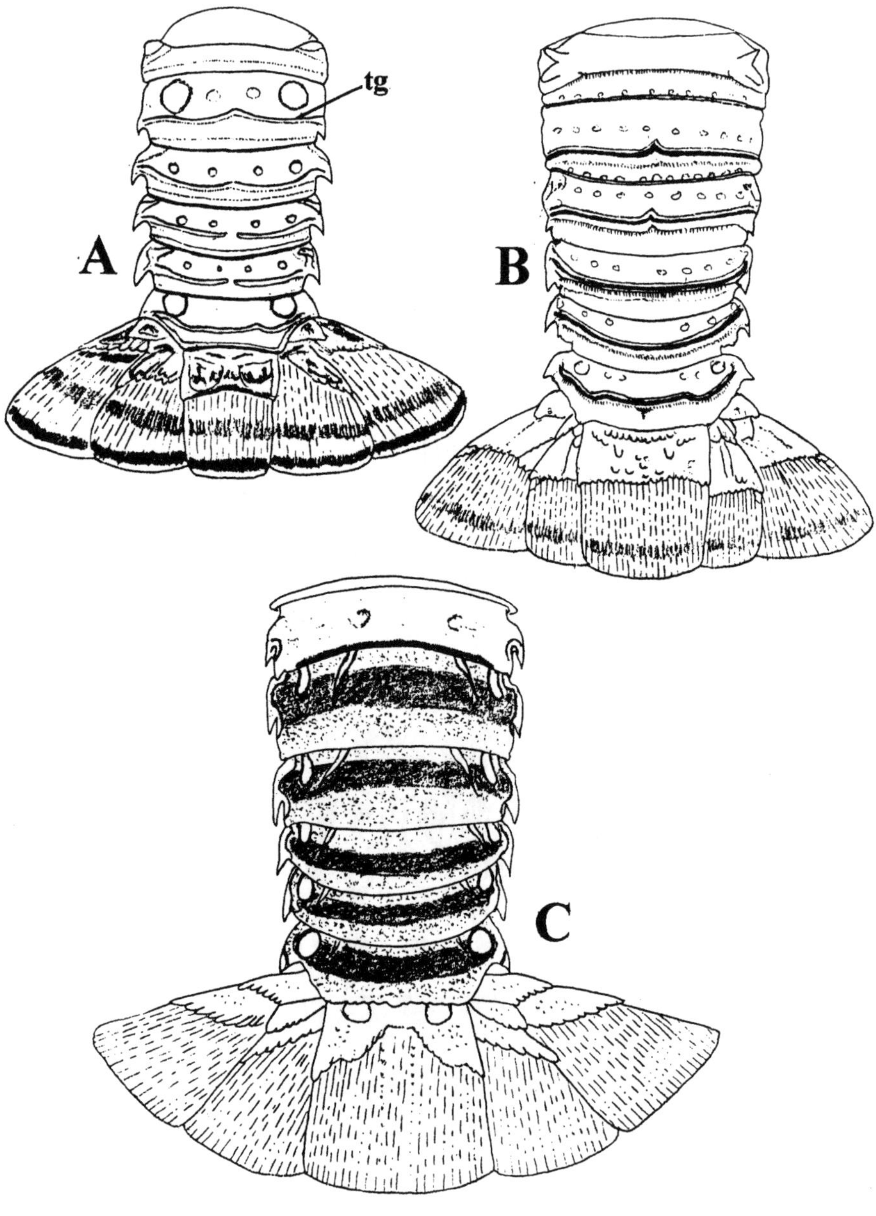

Fig. 11.13 Abdomens from dorsal aspect of spiny crawfishes. A. *Panulirus argus*. B. *Panulirus cygnus*. C. *Panulirus ornatus*. See table 11.3.

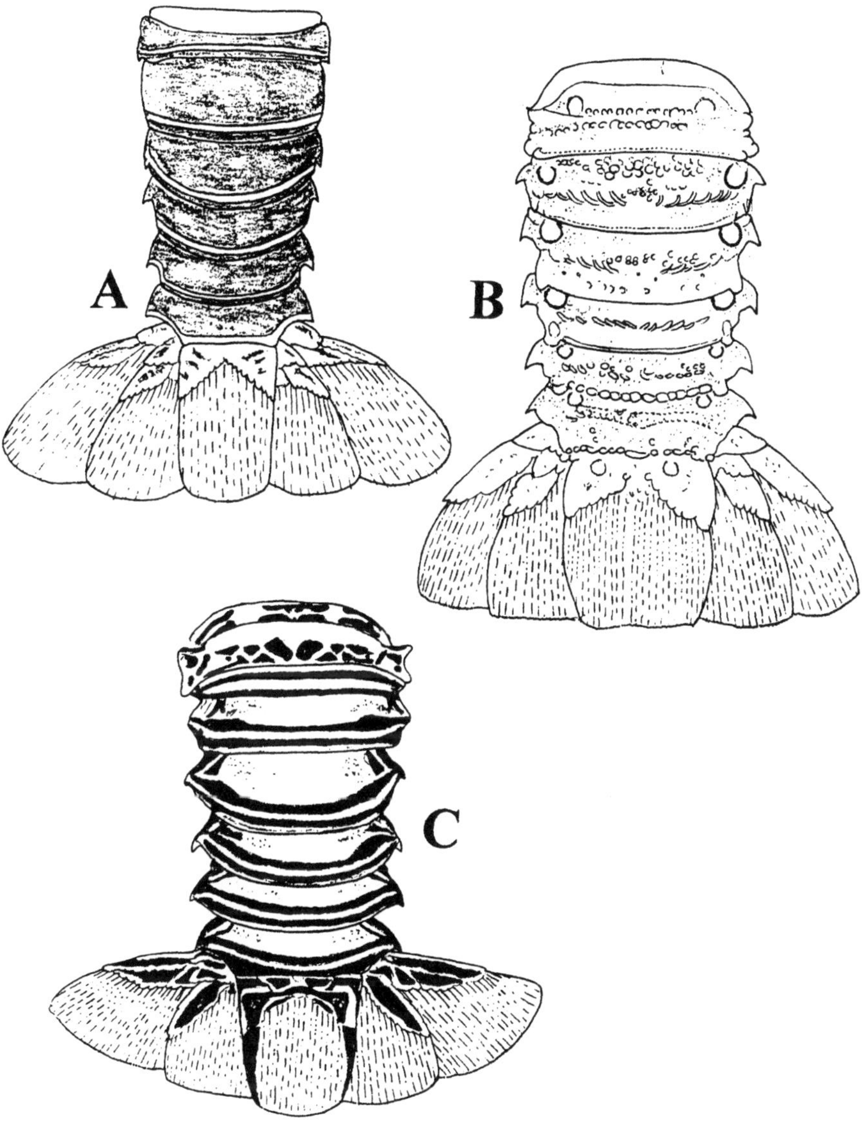

Fig. 11.14 Abdomens from dorsal aspect of spiny crawfishes. A. *Panulirus polyphagus*. B. *Panulirus homarus*. C. *Panulirus versicolor*. See table 11.3.

Table 11.3 Comparative features of spiny crawfishes (*Panulirus*).

SPECIES	*Panulirus argus* Caribbean Spiny Crawfish	*Panulirus cygnus* Australian Spiny Crawfish	*Panulirus ornatus* Ornate Spiny Crawfish
Antennular plate (ap, fig.11.11, A)	with four large spines(fig. 11. 11, A)	with two large spines (fig. 11. 11, B)	with four large spines (fig. 11.11, C)
Transverse groove (tg, fig. 11.13, A) on abdominal segments	present, not crenulated and not interrupted medially (fig. 11. 13, A)	present,not crenulated and not interrupted medially (fig. 11.13, B)	absent (fig. 11.13, C)
Distinctive colour features of abdomen	white/yellow circular patches on segment sides, large on second to sixth, background brownish green	scattered small pale spots, sometimes faint, background pink to dark brown	two oblique yellow-white patches on segment sides, broad dark band across segment middle, background green-grey
Chief fishing regions	Bermuda to Brazil	Western Australia	Indo-West Pacific
Maximum lengths	45 cm	40 cm	50 cm

Panulirus polyphagus Mud Spiny Crawfish	*Panulirus homarus* Scalloped Spiny Crawfish	*Panulirus versicolor* Painted Spiny Crawfish
with two large spines (fig. 11.12, A)	with four large spines (fig. 11.12, B)	with four large spines (fig. 11.12, C)
absent (fig. 11.14,A)	present, crenulated and medially interrupted (fig. 11.14, B)	absent (fig. 11.14, C)
narrow white band near hind margin of segments, background green-grey	minute white spots and large circular patch on second to sixth segments (smaller on fifth and sixth), background brownish red	hind margins of segments with two narrow deep black bands separated by intense white band, background green
Bay of Bengal, Gulf of Thailand	Indian and Taiwan waters	Indo-West Pacific
40 cm	31 cm	40 cm

11.3.5 Rock crawfishes (Genus *Jasus*)

The seven species assigned to this genus all occur in temperate waters of the southern Atlantic and Pacific Oceans. Features of the five species which are most likely to be imported into European and UK markets are compared in table 11.4 and illustrated in figs 11.15, 11.16.

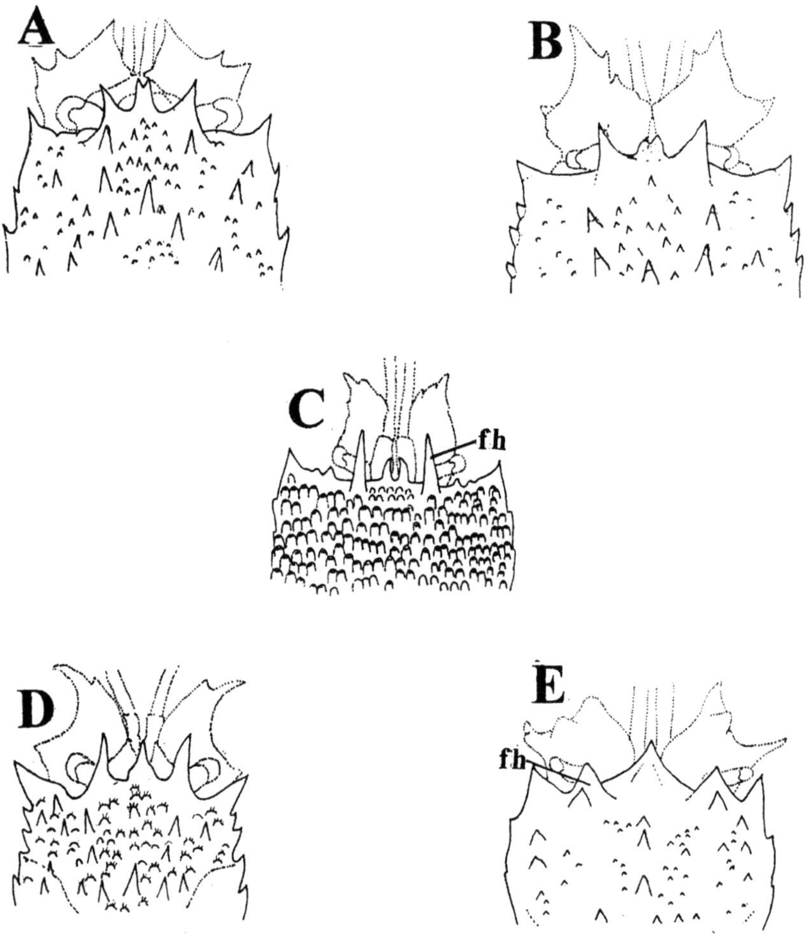

Fig. 11.15 Anterior parts of carapace from dorsal aspects of rock crawfishes. A. *Jasus (Jasus) paulensis,* St. Paul Rock Crawfish. B. *Jasus (Jasus) tristani,* Tristan Rock Crawfish. C. *Jasus (Jasus) lalandii,* Cape Rock Crawfish. D. *J (Jasus) edwardsii,* Red Rock Crawfish. E. *Jasus (Sagmariasus) verreauxi,* Green Rock Crawfish. See table 11.4.

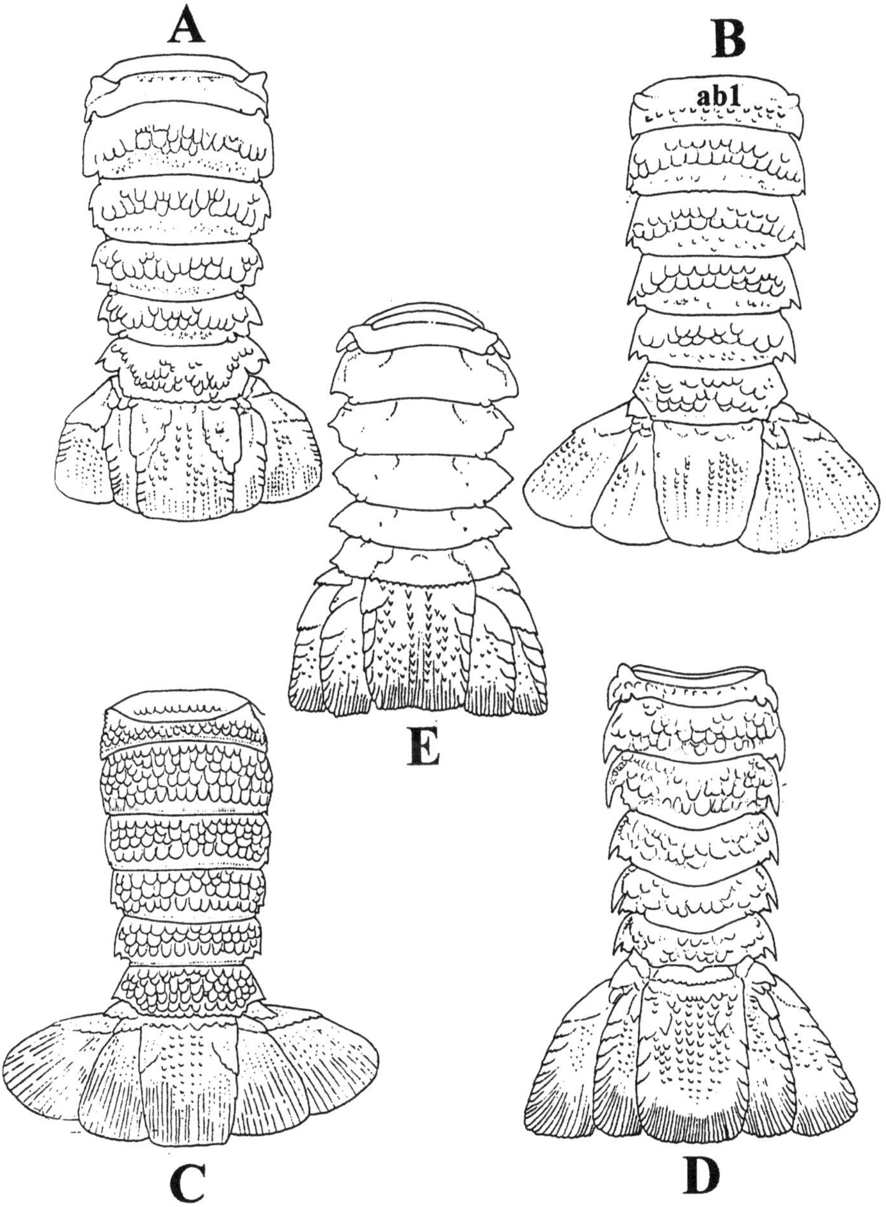

Fig. 11.16 Abdomens from dorsal aspects of rock crawfishes. A. *Jasus* (*Jasus*) *paulensis*. B. *Jasus* (*Jasus*) *tristani*. C. *Jasus* (*Jasus*) *lalandii*. D. *Jasus* (*Jasus*) *edwardsii*. E. *Jasus* (*Sagmariasus*) *verreauxi*. See table 11.4.

Table 11.4 Comparative features of rock crawfishes (*Jasus*)

SPECIES	*Jasus (Jasus) paulensis* St Paul Rock Crawfish (figs 11.15: 11.16, A)	*Jasus (Jasus) tristani* Tristan Rock Crawfish (figs 11.15: 11.16, B)	*Jasus (Jasus) lalandii* Cape Rock Crawfish (figs 11.15: 11.16, C, Pl.16)	*Jasus (Jasus) edwardsii* Red Rock Crawfish (figs 11.15: 11.16, D)	*Jasus (Sagmariasus) verreauxi* Green Rock Crawfish (figs 11.15: 11.16, E)
Carapace dorsal surface	large spines among scattering of small spines	large spines among scattering of small spines	covered with scale-like closely apposed flat tubercles	large spines among numerous small spines, some setosed	large spines among scattering of small spines
Carapace frontal horns (fh, figs. 11.15, C, E)	narrow and diverging, extending to rostra apex	narrow, not diverging, extending to rostra apex or beyond	very narrow, long, not diverging, extending beyond rostral apex	narrow, long, not diverging, extending beyond rostra apex	very broad, short, not extending to rostral apex
Scale-like sculpturing on abdominal segments	present	present	present	present	absent
Dorsal surface of first abdominal segment (ab1, fig. 11.16, B)	with transverse row of scale-like tubercles	with transverse row of scale-like tubercles	with transverse row of scale-like tubercles	with transverse row of scale-like tubercles	without tubercles

SPECIES	*J (J) paulensis* St Paul Rock Crawfish (figs 11.15: 11.16, A)	*J (J) tristani* Tristan Rock Crawfish (figs 11.15: 11.16, B)	*J (J) lalandii* Cape Rock Crawfish (figs 11.15: 11.16, C, Pl. 16)	*J (J) edwardsii* Red Rock Crawfish (figs 11.15: 11.16, D)	*J (S) verreauxi* Green Rock Crawfish (figs 11.15: 11.16, E)
Dorsal surfaces of abdominal segments	with broad smooth anterior region	with broad smooth anterior region	with narrow smooth anterior region	with narrow smooth anterior region	whole surface entirely smooth
Dorsal surfaces of second to sixth abdominal segments	with numerous narrowed scale-like tubercles appearing 'crowded'	with well developed broad scale-like tubercles not appearing 'crowded'	covered with rows of scale-like closely apposed flat tubercles	with few rows of tubercles, some not well defined or uniformly arranged	entirely smooth
Colour of abdominal segments	Reddish purple, smooth regions speckled white	Reddish purple, with white spots	Dark purple with oblique lateral bands of white	Reddish brown with white retic-ulations	greenish to brownish yellow
Fishing regions	St Paul & Amsterdam Isl. Indian Ocean	Tristan da Cunaha group Isl. S. Atlantic	S. W. Africa	New Zealand, including Chatham Isl.	S. E. Aus-tralia, N. Island, New Zealand
Maximum lengths	37 cm	35.5 cm (males) 27 cm (females)	46 cm	58 cm (males) 43 cm (females)	60 cm

11.3.6 Whip crawfishes (Genus *Puerulus*)

The four known species of whip crawfishes all occur in Indo-Pacific waters of moderate depth. To date the frozen tails of only the Arabian Whip Crawfish (*Puerulus sewelli,*) have been imported into the UK. The 'Aden Tails' marketed are probably of this species.

Features that distinguish whip crawfishes from other crawfishes are listed in table 11.2 (p. 202). The Arabian Whip Crawfish is illustrated in figs.11.9, D, 11.10, D, 11.17 & Pl. 19.

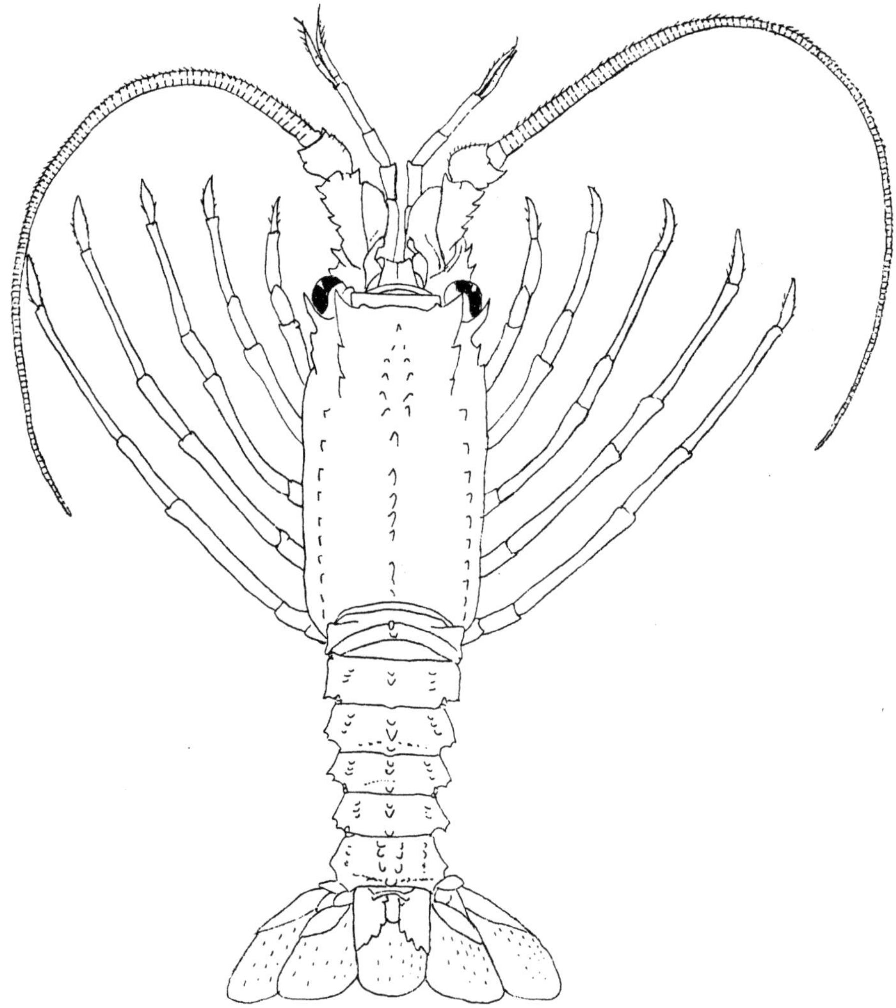

Fig. 11.17 *Puerulus sewelli*, The Arabian Whip Crawfish, 20 cm total length.

11.4 SHOVEL NOSED LOBSTERS

11.4.1 Imported species

Many species of shovel-nosed lobsters (family Scyllaridae) form incidental catches in traps and nets set to catch crawfishes and are often sold in local markets where they may command quite high prices. It is anticipated that from time to time various species of shovel-nosed lobsters may be offered for sale in the UK and Europe as opportunistic imports since there are no regular commercial fisheries for these lobsters. These imported species will belong to one or more of seven genera: *Evibacus* and *Ibacus* (Fan Lobsters), *Arctides* (Spanish Lobsters), *Parribacus* (Mitten Lobsters), *Scyllarides* (Slipper Lobsters), *Scyllarus* (Locust Lobsters) and *Thenus* (Flathead Lobsters). These lobsters can be identified to the relevant genus from the features listed in table 11.5 and figs 11.18, 11.19.
To the author's knowledge at least three species have been occasionally imported. The colourful Royal Spanish Lobster (*Arctides regalis*) has been offered for sale by dealers in marine aquarium fishes. Mounted specimens of the Sculptured Mitten Lobster (*Parribacus antarcticus*) are brought back by visitors returning from Thailand where they are sold as tourists trophies. The Moreton Bay Flathead Lobster (*Thenus orientalis*) is imported frozen into the UK by restaurateurs.

11.4.2 Distinguished from native species

The clearly defined elongate configuration of lobules on the mid-line dorsal surface of the abdominal segments and the somewhat blunt teeth of the distal margin of the antennae (see fig. 6.5).are features that distinguish the native Small European Locust Lobster (*Scyllarus arctus*) and Pigmy Locust Lobster (*S. pygmaeus*) from imported species.

The following combined features will enable the Mediterranean Locust Lobster (*Scyllarides latus*) to be recognised from imported species: (i) posterior margin of second abdominal segment with a large posterior tooth; (ii) carpus of first pereiopod with dorsal groove; (iii) medial portion of fourth abdominal segment not raised; (iv) outer two large circular reddish spots on first abdominal segment with inner margins concave (fig. 6.4).

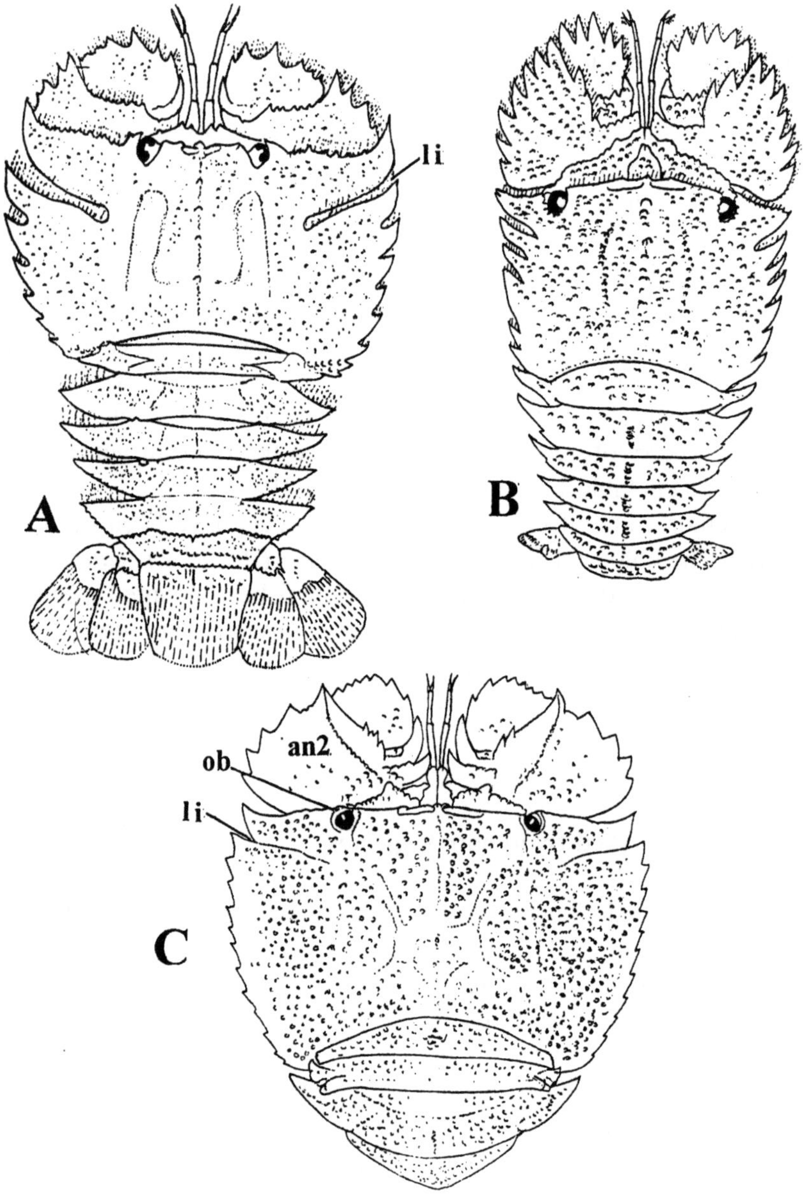

Fig. 11.18 Carapace and abdomens, from dorsal aspect, of three genera of shovel-nosed lobsters. A. *Ibacus novemdentatus*, Smooth Fan Lobster, 19 cm. B. *Parribacus antarcticus*, Sculptured Mitten Lobster, abdomen shown slightly folded, 20 cm. C. *Evibacus princeps*, Shield Fan Lobster, abdomen shown folded, 33 cm. Measurements are total lengths.

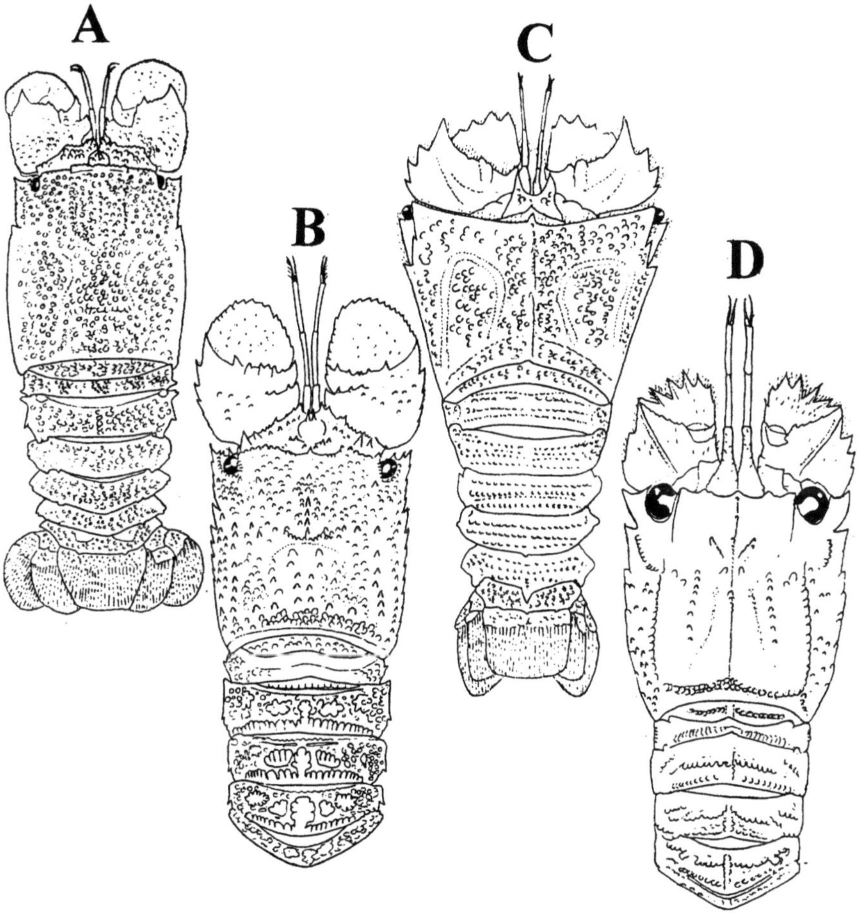

Fig. 11.19 Carapace and abdomens, from dorsal aspect, of four genera of shovel-nosed lobsters. A. *Scyllarides herklotsii*, Red Slipper Lobster, 32 cm. B. *Arctides regalis*, Royal Spanish Lobster, distal part of abdomen folded, 17 cm. C. *Thenus orientalis*, Moreton Bay Flathead Lobster, 25 cm. D. *Scyllarus batei*, Soft Locust Lobster, distal part of abdomen folded, 7.0 cm. Measurements are total lengths.

Table 11.5 Comparative features of the seven genera of shovel-nosed lobsters.

SPECIES	*Ibacus* Fan Lobsters (fig. 11.18, A)	*Parribacus* Mitten Lobsters (fig. 11.18, B)	*Evibacus* Fan Lobsters (fig. 11.18, C)
Carapace dorsal surface	noticeably flatt-ened	noticeably flattened	noticeably flatt-ened
Carapace lateral incision (li, figs 11.18, A, C)	deep and open	deep and open	deep and closed
Orbits (ob, fig. 11.18, C)	with anterior gap	with anterior gap	without anterior gap
Position of eyes	very far from carapace margins	far from carapace margins	very far from carapace margins
Distance between orbits	narrow	wide	wide
Dorsal surface of abdomen	smooth or minu-tely pitted	with large flat tubercles	finely granular
Margins of second antenna (an2, fig 11.18, C)	with numerous teeth	with numerous teeth	with numerous teeth
Flagellum on exopod of third maxilliped (see fig. 1.2, mxp3, ex, fl)	present	present	present

Scyllarides Slipper Lobsters (fig. 11.19, A)	*Arctides* Spanish Lobsters (fig. 11.19, B)	*Thenus* Flathead Lobsters (fig. 11.19, C)	*Scyllarus* Locust Lobsters (fig. 11.19, D)
vaulted	vaulted	vaulted	vaulted
very shallow	very shallow	very shallow	very shallow
with anterior gap	with anterior gap	with anterior gap	with anterior gap
very near carapace margins	very near carapace margins	at carapace margins	very near carapace margins
wide	wide	very wide	wide
with large flat tubercles	sculptured	tubercles arranged in transverse ridges	sculptured
without or with few teeth	with numerous teeth	with numerous teeth	with numerous teeth
present	present	absent	absent

11.5 SQUAT LOBSTERS

11.5.1 Imported species

Two species of squat lobsters have been imported from time to time into UK and probably into European markets. *Cervimunida johni* (fig. 11.20, A, Pl. 20), the Yellow Squat Lobster or Langostino Amarillo and *Pleuroncodes monodon* (fig. 11.20, B, Pl. 21), the Red Squat Lobster or Langostino Colorado. The chief fishery grounds for these species are off the Peruvian and Chilean coast. The vernacular name 'Red Lobster' or 'Red Crab' has been used also for *Pleuroncodes planipes*, a species that may be identical with *P. monodon* and occurring off the Californian coast but for which there is no commercial fishery.

These squat lobsters are better known in the USA where they are usually marketed as 'langostino'. Confusion can arise when they are imported into the UK or Europe under this name since langostino is the Spanish name for shrimp.

The Yellow Squat Lobster can be recognised by rows of minute dorsal spines (**sp**, fig.11.20, A) on the abdominal segments and the Red Squat Lobster by the presence of an extra pair of spines (**sp**, fig. 11,20, B) near the rostral base.

The total length of these squat lobsters (average 5-6 cm) is smaller than the majority of crawfishes and their economic value, similar to nephropid lobsters, depends upon the enormous numbers fished because the meat is limited only to the abdomen or 'tails'.

11.5.2 Distinguished from native species

Unlike native squat lobsters (pp 102-107) these imported species have the conspicuous striate sides of the carapace clearly visible when viewed from dorsal aspect as shown in the figures.

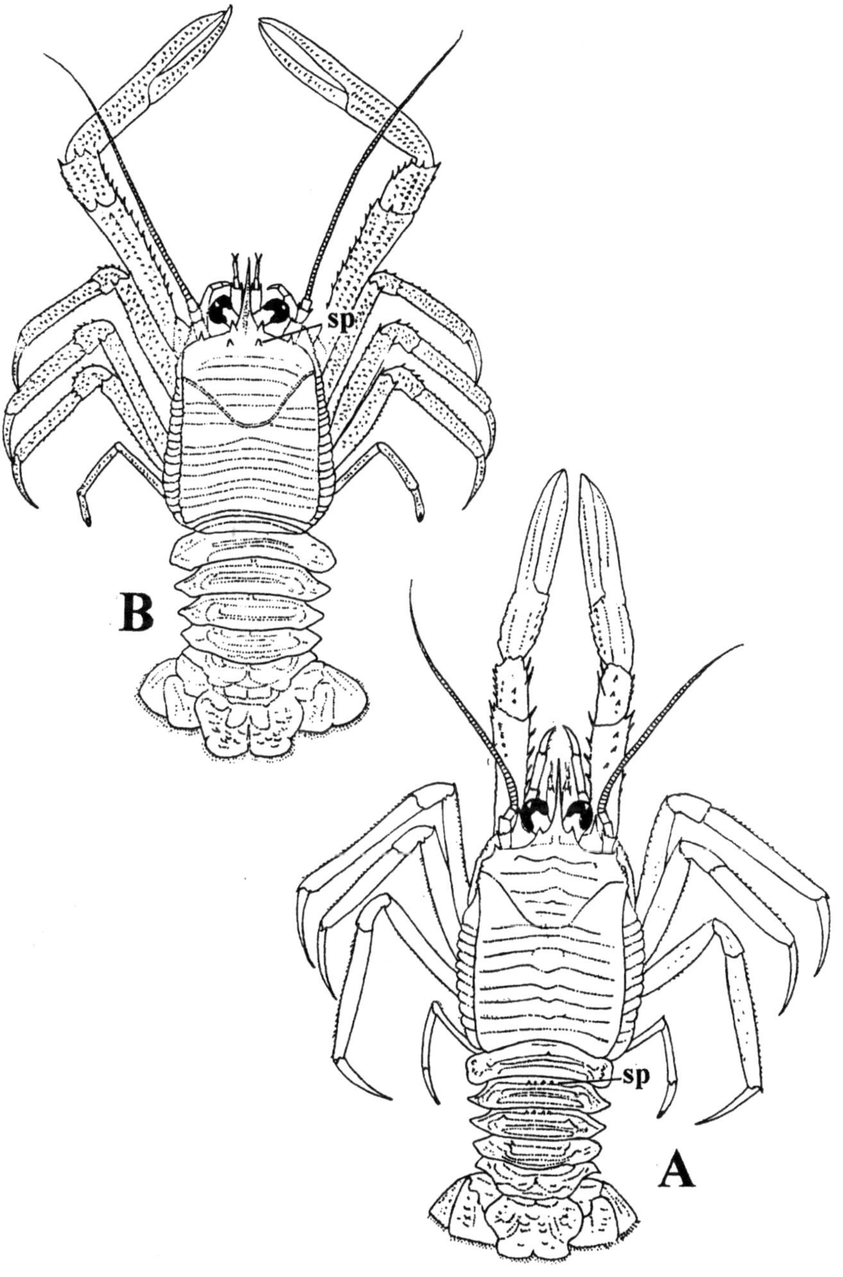

Fig. 11.20 A. *Cervimunida johni*, Yellow Squat Lobster. B. *Pleuroncodes monodon*, Red Squat Lobster. For explanation of annotations see opposite page.

11.6 STONE CRABS

11.6.1 Imported species

Whole stone crabs are rarely imported. Meat extracted from the chelipeds [claws], walking legs [pereiopods] and 'tail' [abdomen] is packed in various combinations (sometimes shredded) and marketed frozen or tinned. The meat contained in individual products may be that of more than one species and meat from the Patagonian Royal Stone Crab has been sold under the name of 'Alaskan King Crab' (see below).

Meat from at least two species of stone crabs has been marketed from time to time in the UK and European countries.

The Patagonian Royal Stone Crab, *Lithodes santolla* (fig. 11.21, Pl. 22) is often cited under an older name of *Lithodes antarctica*. The species is fished in shallow waters of southernmost South America and can be recognised by the conspicuous trifid rostrum (fig. C) and the smooth branchial regions (**br**, fig. B) armed with less than 19 spines. Two forms of this species, both of a similar size, are often caught in the same hauls: a very spiny form (shown in fig. A), and a less spiny form (fig. B). The largest specimens marketed of this crab are in excess of 19.8 cm carapace length.

The Alaskan King Stone Crab, *Paralithodes camtschaticus* (fig. 11.22, A, Pl. 23) is fished in deep and shallow waters of the Sea of Japan to northern British Columbia. The reddish sub-quadrate carapace (**cr**, fig. A) and stout broad second (**pr2**) to fourth (**pr4**) pereiopods are covered with small and relatively short spines. The largest specimens measure about 21.5 cm carapace length. In the past whole or segments (fig. B) of frozen limbs were occasionally sold in the UK as 'crab sticks'(not to be confused with the products now marketed under this name and composed of fish and crab derivatives), but such crab limbs are now more commonly seen in the USA and Japan when available. The limbs of the Alaskan King Stone Crab are much broader and have shorter spines than those of the Patagonian Royal Stone Crab.

The extensive north Pacific fisheries for the Alaskan King Stone Crab virtually collapsed during the last decade due chiefly to over exploitation. The meat genuinely derived from this crab is unlikely to be seen in any quantity in markets for some time but meat extracted from the two species mentioned below may be occasionally offered as substitutes for that of the Alaskan King Stone Crab.

The Brown King Stone Crab, *Paralithodes brevipes*. This species is fished in shallow waters of the northern Sea of Japan to the Bering Strait. The crab is dark reddish to purple, has robust short walking legs and the stout rostrum is bent downwards and has a rounded tip.

The Blue King Stone Crab, *Paralithodes platypus*, is fished in deep waters of the Sea of Japan to Alaska. The colour is bluish purple, the carapace is distinctly rounded in outline, there are fewer tubercles/spines on the carapace and the rostrum is small and short.

The vernacular name 'King Stone Crab' should be restricted to species belonging only to the genus *Paralithodes*.

11.6.2 Distinguished from native species

The native Northern Stone Crab, *Lithodes maja* (p. 96) can be easily distinguished from imported Patagonian Royal Stone Crabs by the bifid rostrum (see fig. 8.7, B) and from Alaskan Brown and Blue King Stone Crabs by the very spiny and distinctly oval carapace, the long rostrum and the spinosed long thin walking legs [second to fifth pereiopods].

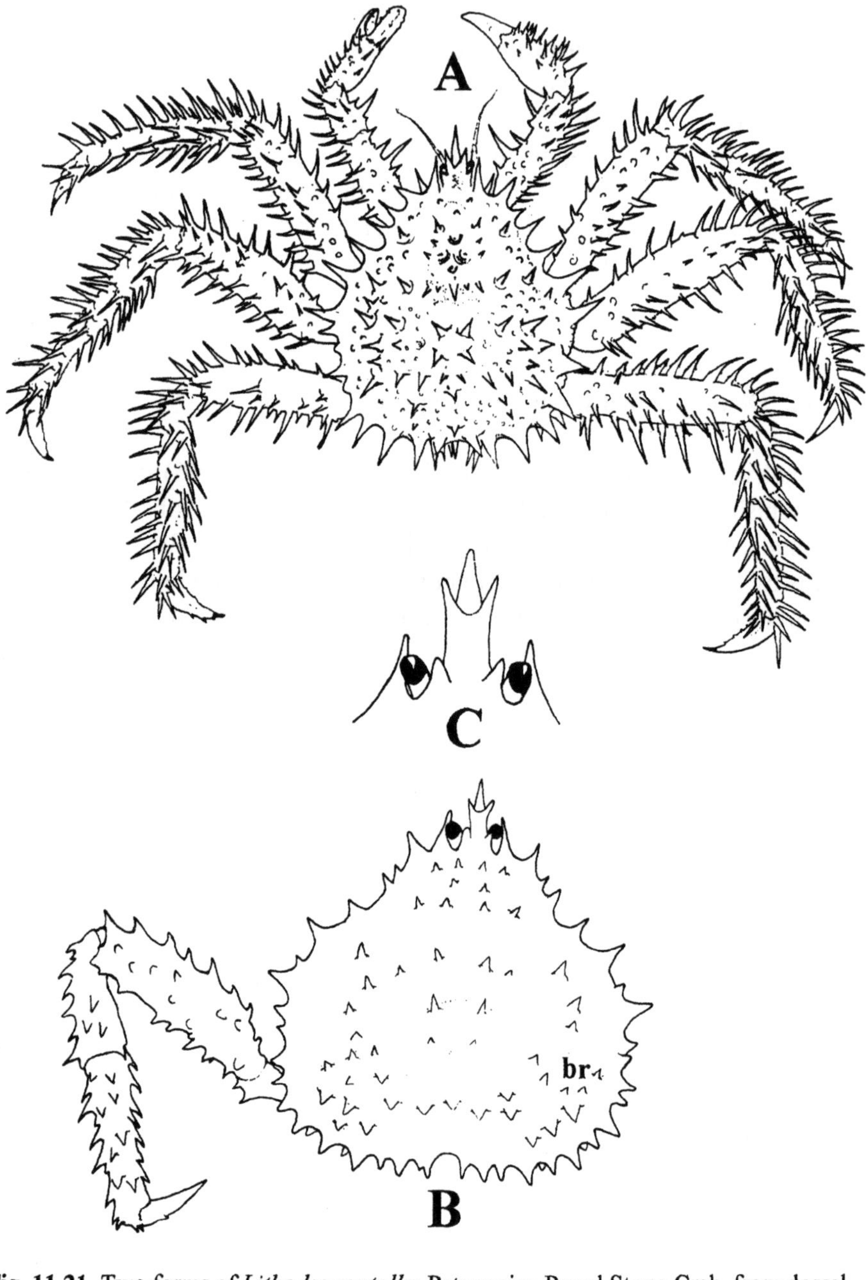

Fig. 11.21 Two forms of *Lithodes santolla*, Patagonian Royal Stone Crab, from dorsal aspects and of similar sizes, both measuring 7 cm carapace length. A. Spiny form, whole crab. B. Less spiny form, carapace and fourth left pereiopod shown. C. Rostrum of latter. For explanation of annotation see p. 222.

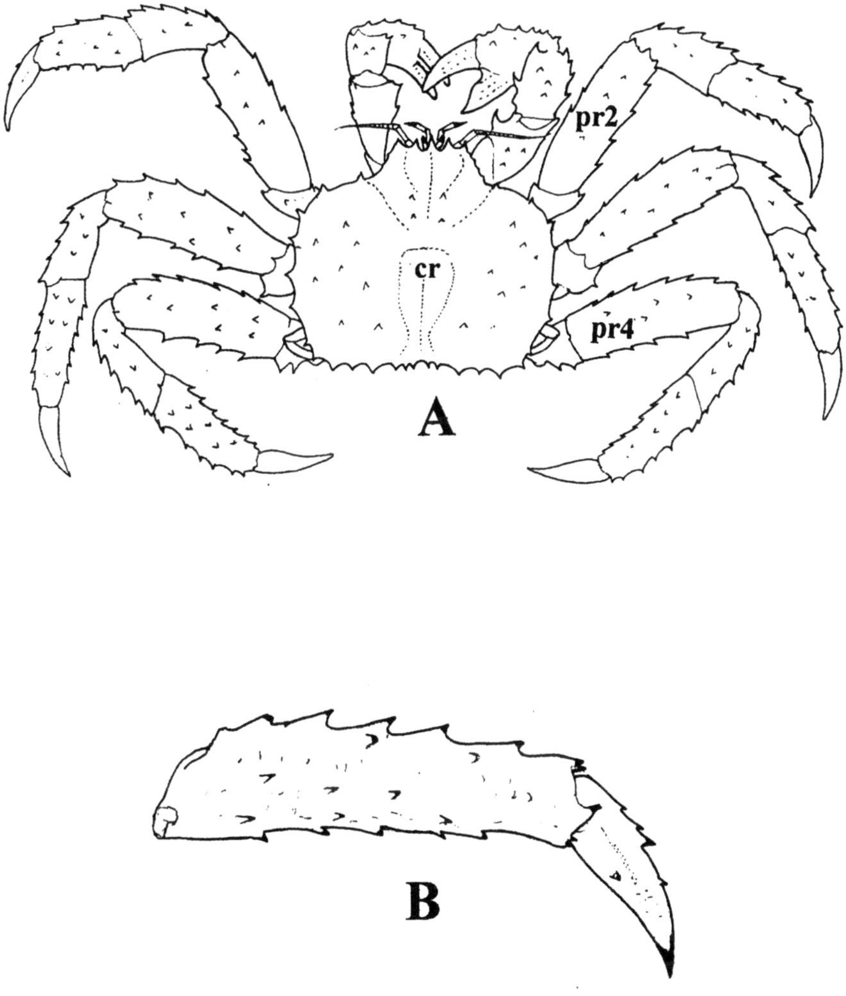

Fig. 11.22 A. *Paralithodes camtschaticus*, Alaskan King Stone Crab. B. Propodus and dactylus of third right pereiopod from dorsal aspect. For explanation of annotations see p. 222.

11.7 TRUE CRABS

11.7.1 Imported species

A considerable number of true crabs are fished for human consumption by Indo-Pacific communities but, when compared with crawfishes, only a few species seem to have been commercially imported into European and UK markets.

The West African Geryonid Crab, *Chaceon maritae* (fig. 11.23, Pl. 32) is fished off the west coast of Africa. The claws (fig. C) sold in European markets are from crabs caught mainly by Spanish trawlers.

The Mud Swimming Crab, *Scylla serrata* (fig. 11.24, A, Pl. 28) is fished in coastal waters in several regions throughout its range in the Indo-Pacific and often in estuaries. Crabs are occasionally seen for sale in a few of the more exclusive UK and European food halls. This crab is also known as the Mangrove Crab, Giant Mangrove Swimming Crab, Swamp Crab and Stone Crab. Local names are: Chimbori, Khadapi Chimbori (Bombay region); Nona Kankra (Bengal region), Alimango (Philippines), Ch'inghsieh (China) and Samoan Crab (Samoa). There are several colour varieties of this crab and some authorities have suggested that at least three species and one sub-species may be represented.

The Blue Swimming Crab, *Portunus pelagicus* (fig. 11.24, B, Pl. 27) is the most important commercial species of the various swimming crabs fished in Indo-Pacific coastal waters by many local communities. Some crabs sold in European markets may originate from the Israel coast. There, and in the Suez Canal, the species has become established. Alternative vernacular names used for this crab are: Blue Manna, Bluey, Sand Crab (Australia); Radjungan (Malaysia) and Nili Chimbori (India).

The Mitten Crab (*Eriocheir sinensis*) is occasionally offered for sale in European fish markets catering for oriental communities. It is an introduced species and is discussed on p.162.

The Atlantic Snow Spider Crab, *Chionoecetes opilio opilio* (fig. 11.25, Pl. 34) supports trap fisheries in the Gulf of St Lawrence and Newfoundland and also in the North Pacific. Some authorities regard the Pacific populations as a distinct sub-species (Pacific Snow Spider Crab, *Chionoecetes opilio elongatus*) or species (*Chionoecetes elongatus*). The closely related Snow Queen Spider Crab *Chionoecetes bairdi* and the Tanner Spider Crab *Chionoecetes tanneri* are also commercially fished in Northern Pacific regions. All three (or four) species probably have been imported into UK and European markets from time to time.

11.7.2 Distinguished from native species

It is unlikely that specimens or claws of the West African Geryonid Crab will be confused with the native Red Geryonid Crab (p. 138) since there is no commercial fishery for this latter species at present. The imported African species is without a small hump over each orbital surface and the dactyl of the fifth pereiopod is dorsally compressed, ie. the outer margin (**om**, fig. 11.23, B) is broadened. The chelipeds (fig. C) are also much shorter and more curved that those of the native Red Geryonid Crab.

The nine equal size carapace antero-lateral teeth (fig. 11.24, A) distinguishes the Mud Swimming Crab from all native swimming crabs that have far fewer teeth.

The blue coloured carapace of imported Blue Swimming Crabs has a conspicuous white reticulated pattern and the distal margin of the cheliped merus (**mrs**, fig. 11.24, B) is armed with conspicuous spines. These two features distinguish this species from locally caught or imported Chesapeake Bay Crabs (see fig. 10.15, p. 148).

The spines on the carapace of snow spider crabs are inconspicuous, the antero-lateral margins (**al**, fig. 11.25) inflated and the rostrum (**rs**) is short and flat. These features distinguish snow spider crabs from the Thornback Spider Crab (fig. 10.27, p. 172), the only native species of comparable size with which it could be confused.

The Atlantic and Pacific Snow Spider Crabs have a carapace that is as broad as long. The carapace is broader than long for the Snow Queen Spider Crab and Tanner Spider Crab. The latter has the branchial regions noticeably swollen and the rostrum is slightly upturned.

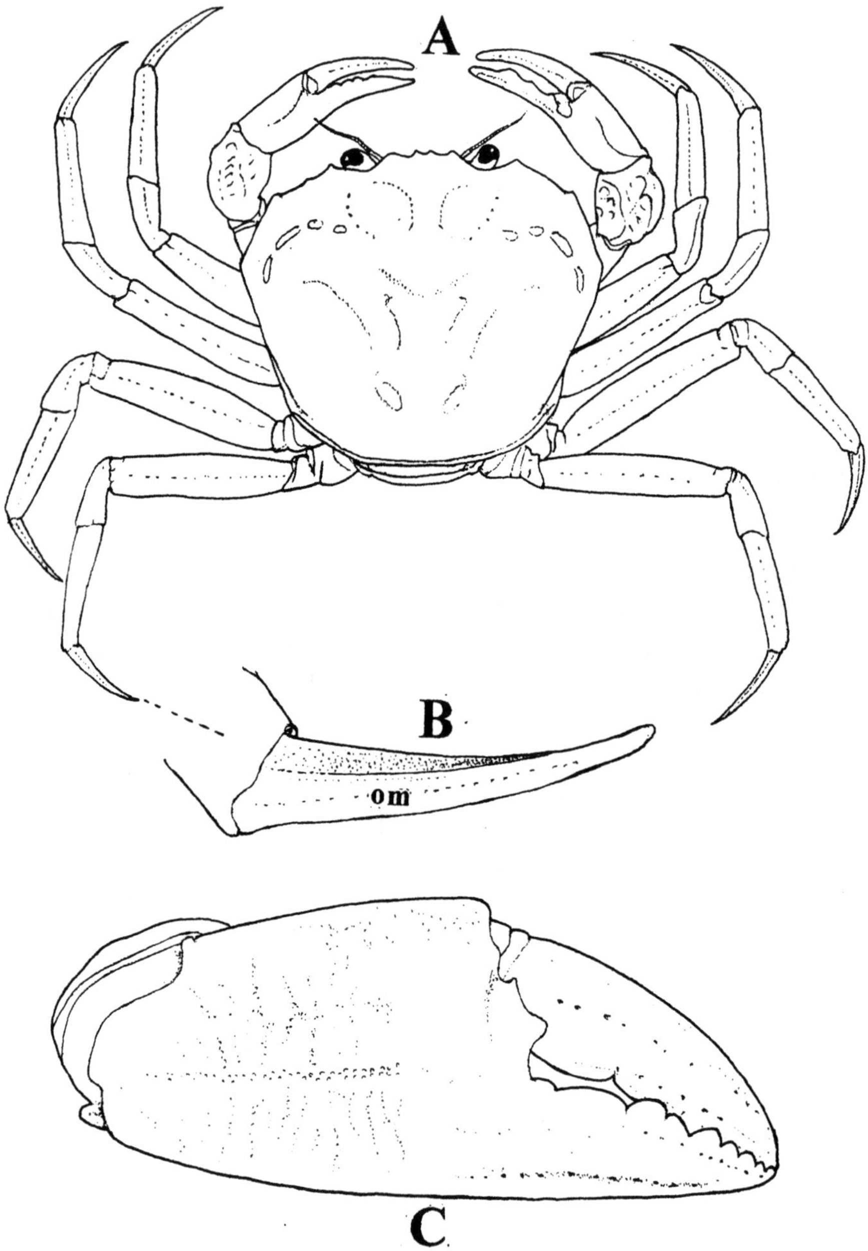

Fig. 11.23 *Chaceon maritae*, West African Geryonid Crab. A. Dorsal aspect of whole crab, 14 cm carapace length. B. Dactylus of fifth left pereiopod from outer aspect. C. Right cheliped from outer aspect. For explanation annotation see p.227.

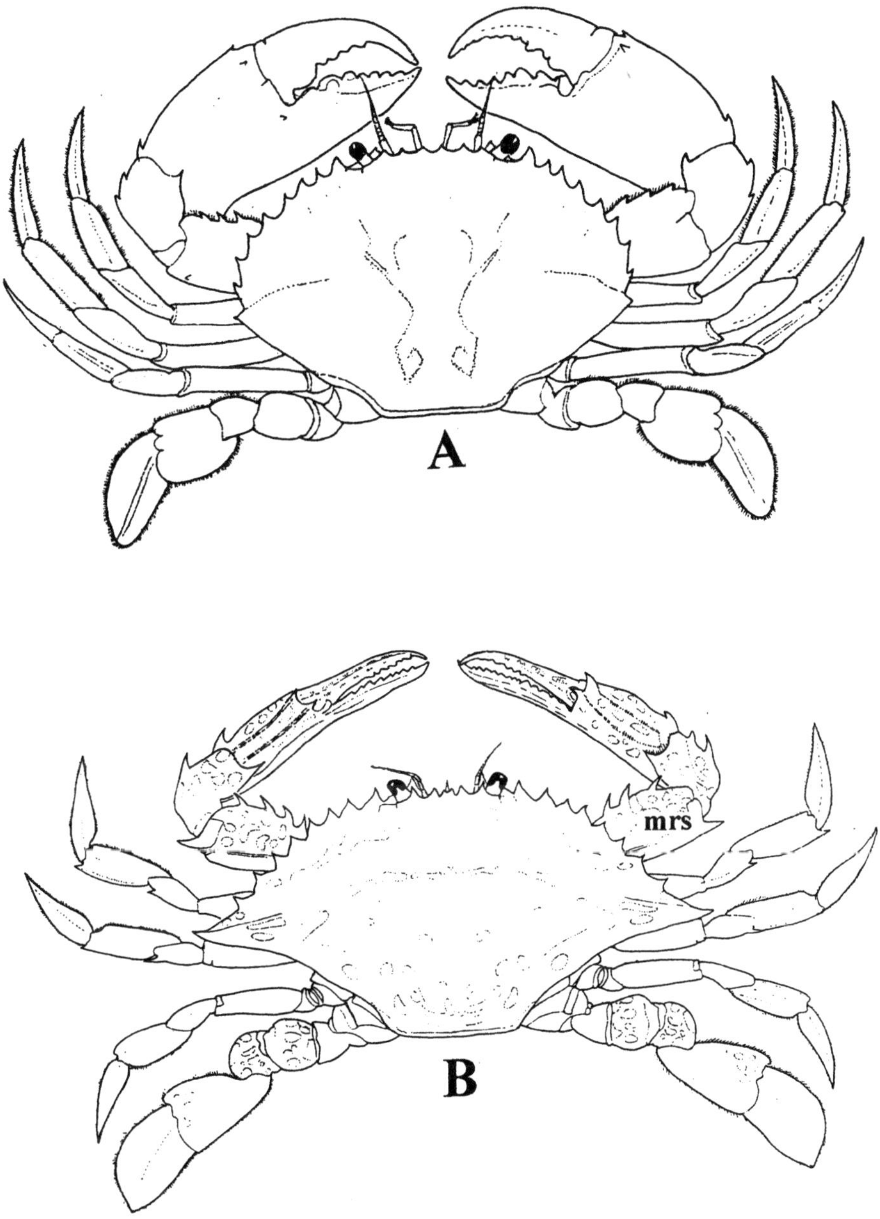

Fig. 11.24 A. *Scylla serrata*, Mud Swimming Crab, 16 cm approx. maximum carapace length. B. *Portunus pelagicus*, Blue Swimming Crab, 13 cm approx. maximum carapace length. For explanation of annotation see p.227.

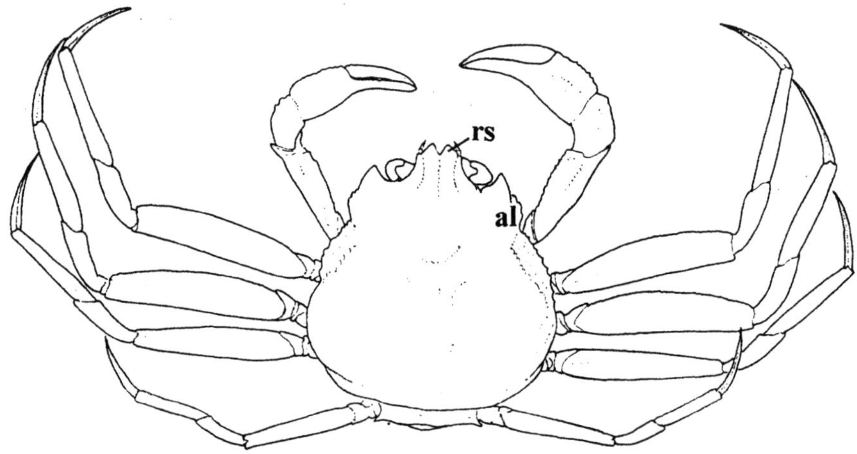

Fig. 11.25 Atlantic Snow Spider Crab, *Chionoecetes opilio opilio*, 12.5 cm carapace length. For explanation of annotations see p. 227.

FURTHER READING

Anon. *Crayfish Alert-Keeping the lid on crayfish (information on legal aspects of traded crayfishes in Britain.* MAFF Publications London, SE99 7TP.

Dawson, E. W. (1989) King Crabs of the World (Crustacea: Lithodidae) and their fisheries. *Miscellaneous Publication* **101,** *New Zealand Oceanographic Institute.*

Dore, I. (1989) *The New Frozen Seafood Handbook. A Complete Reference for the Seafood Business.* A Osprey Book, Van Nostrand Reinhold, New York (crabs, crayfishes and crawfishes pp 52-65, lobsters, pp 146-157).

Guinot, D. (1967) *Les crabes comestables de l'Indo-Pacific.* Fondation Singer-Polignac, Paris.

Herrick, F. H. (1895) The American Lobster: a study of its habits and development. *Bulletin of the United States Fish Commission*, **15**, 1-252.

Holthuis, L. B. (1991) Volume 13, Marine Lobsters of the World. FAO Species Catalogue. *FAO Fishery Synopsis*, **125**,.FAO Rome.

McLeese, D. W. and Wilder, D. G. (1964) Lobster storage and shipment. *Bulletin of the Fisheries Research Board of Canada*, **147**, 1-69.

Merrick, J. R. (1993) Freshwater Crayfishes of New South Wales. Linnean Society of New South Wales. Marrickville, NSW.

Taylor, H. (1975) *The Lobster: its life cycle.* Sterling Publishing Company Incorporated, New York.

Williams, A. B. (1988) *Lobsters of the World-an illustrated guide. Lobsters of the World in US Trade.* Osprey Books, New York and Scandinavian Fishing Year Book, Denmark.

Colour plates

Crayfishes, lobsters and crabs traded in the UK and Europe

Plate 1 *Austropotamobius pallipes*, White-clawed Crayfish, Hertfordshire, UK. Male. 4.2 cm total length.

Plate 2 *Astacus astacus*, Noble Crayfish, River Elbe, Germany. Male. 5.5 cm total length.

Plate 3 *Astacus leptodactylus leptodactylus*, Long-clawed Crayfish, Grand Union Canal, London, UK. Male. 14.5 cm total length.

Plate 4 *Pacifastacus (P) leniusculus leniusculus*, Signal Crayfish, Washington County, USA. Male. 15 cm total length.

Plate 5 *Orconectes (F) limosus*, Striped Crayfish Virginia, USA. Female. 7 cm total length.

Plate 6 *Procambarus (S) clarkii*, Red Swamp Crayfish, Harrods Food Hall, London, UK. Male. 12 cm total length.

Plate 7 *Cherax destructor*, Melbourne Yabby, Hermannsburg, South Australia. Female. 11.5 cm total length

Plate 8 *Cherax tenuimanus*, Marron Harvey River, Western Australia Female. 10 cm total length.

Plate 9 *Homarus gammarus*, European Clawed Lobster, Tunisian market. Male. 26 cm total length.

Plate 10 *Homarus americanus*, American Clawed Lobster, South Devon, UK. Male. 26 cm total length (Probably escaped from captivity as indicated by banded claw).

Plate 11 *Nephrops norvegicus*, Norway Clawed Lobster, Loch Torridon, Scotland, UK. Male. 12.5 cm total length.

Plate 12 *Metanephrops andamanicus*, Andaman Clawed Lobster, Natal, South Africa. Male. 18 cm total length.

Plate 13 *Palinurus elephas*, European Spiny Crawfish, Tunisian market. Male. 23 cm total length.

Plate 14 *Palinurus mauritanicus*, Pink Spiny Crawfish, West Africa. Male. 13 cm total length

Plate 15 *Panulirus regius*, Royal Spiny Crawfish, Senegal, West Africa. Female. 15 cm total length.

Plate 16 *Jasus lalandii*, Cape Rock Crawfish, New Zealand. Male. 21 cm total length.

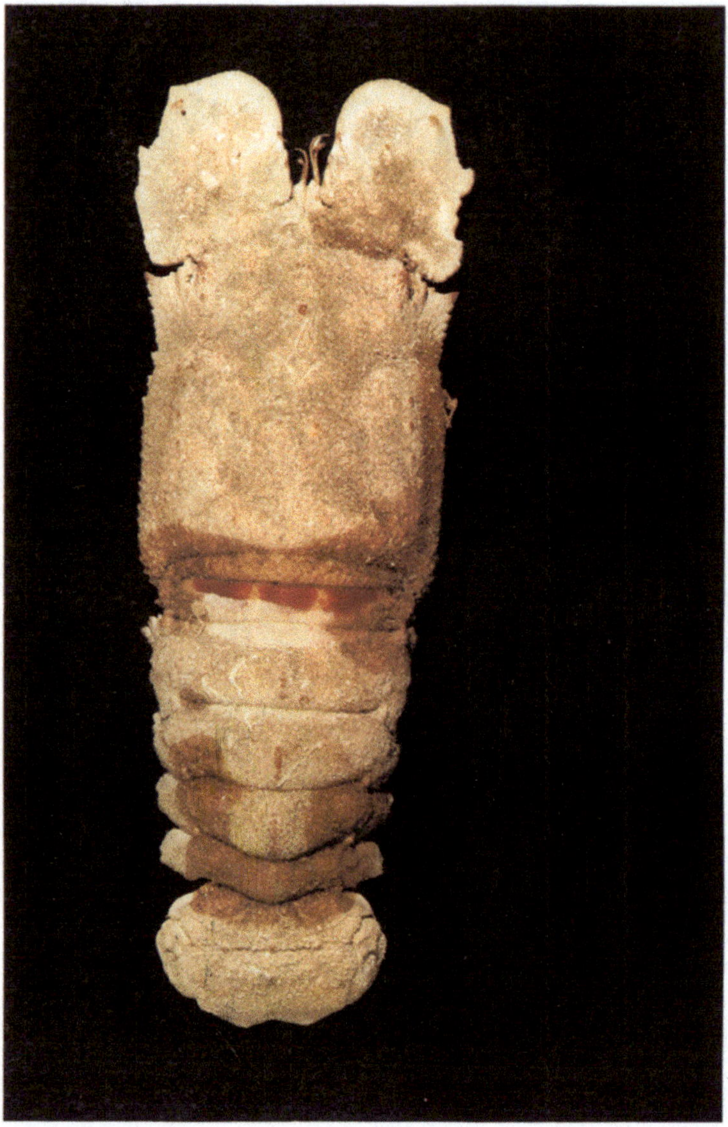

Plate 17 *Scyllarides latus*, Mediterranean Slipper Lobster, Madeira. Female. 17 cm total length.

Plate 18 *Scyllarus arctus*, Small European Locust Lobster, Guernsey, Channel Islands, UK. Male. 10 cm total length.

Plate 19 *Puerulus sewelli*, Arabian Whip Crawfish, Arabian Gulf. Male. 14 cm total length.

Plate 20 *Cervimunida johni*, Yellow Squat Lobster, Chile, South America. Male. 10 cm total length.

Plate 21 *Pleuroncodes monodon*, Red Squat Lobster, Panama. Male. 8 cm total length.

Plate 22 *Lithodes santolla*, Patagonian Royal Stone Crab, South Magellan, South America. Female. 5.5 cm carapace length

Plate 23 *Paralithodes camtschaticus*, Alaskan King Stone Crab, Alaska. Female. 16 cm carapace length

Plate 24 *Cancer pagurus*, European Edible Crab, Devon, UK. Male. 8 cm carapace length

Plate 25 *Necora puber*, Velvet Swimming Crab, South coast of UK. Male. 5.5 cm carapace length.

Plate 26 *Callinectes sapidus*, Chesapeake Bay Swimming Crab, Brazil, South America. Male. 8 cm carapace length

Plate 27 *Portunus pelagicus*, Blue Swimming Crab, Gulf of Suez. Male. 9 cm carapace length.

Plate 28 *Scylla serrata*, Mud Swimming Crab, India. Male. 11 cm carapace length.

Plate 29 *Eriphia verrucosa*, Warty Xanthid Crab, Portugal. Female. 4.3 cm carapace length.

Plate 30 *Eriocheir sinensis*, Mitten Crab, Middlesex, UK. Male. 5 cm carapace length

Plate 31 *Uca tangeri*, Moroccan Fiddler Crab, Ria Formosa, Portugal. Male. 2.3 cm carapace length.

Plate 32 *Chaceon maritae*, West African Geryonid Crab, South Atlantic. Male. 14 cm carapace length.

Plate 33 *Maja squinado*, Thorn-back Spider Cra, Devon, UK. Female. 14 cm carapace length

Plate 34 *Chionoecetes opilio opilio*, Atlantic Snow Spider Crab, Canadian Atlantic. Female. 12 cm carapace length.

Index

(pages with illustrations are shown in bold)

Abbreviations x
Abdomen 1-3
Aden tails 214
Aegean freshwater crab,
see *Potamon potamios*
Aetekrabbe 158
African clawed lobster,
see *Metanephrops mozambicus*
Alaskan king stone crab,
see *Paralithodes camtschaticus*
Alimango 226
American
clawed lobster,
see *Homarus americanus*
river crayfish 182
Amerikanischer flusskrebs 182
Anacra 146
Andaman clawed lobster,
see *Metanephrops andamanicus*
Angular crabs
classification 18
Anomura
taxonomic features 22-3
Antennae 1
Arabian whip crawfish,
see *Puerulus sewelli*
Aragosta mediterranea 66
Araignée
à rostre pointu 174
de mer 172
naine 172
Araña 172
de mar 168
de mar patas largas 170
Arctides
recognition features 219
classification 15
regalis **217**
commerce 215
Armoured clawed lobster,
see *Metanephrops armatus*
Asil pavurya 136
Astacoidea
vernacular name 12
Astacopsis gouldi
classification 14
size 188
Astacus
fluviatilis 36
leniusculus 180
nobilis 36
pallipes 34
Astacus
recognition features 33, 36, 38-39
species classification 13-14
astacus astacus **37, 235**
leptodactylus leptodactylus **39, 236**
salinus **39**
pachypus 33
Astacus (Astacus) astacus 36
Astakos 56, 66
Astakoudaki 72
megalo 70
Astice 56
Atelecyclus
classification 17
recognition features 134
rotundatus **135**
undecimdentatus 134
Atelecyclus septemdentatus 134
Atlantic
pink spiny crawfish,
see *Palinurus mauritanicus*
snow spider crab,
see *Chionecetes opilio opilio*
Augustinus krab 96
Australian spiny crawfish,
see *Panulirus cygnus*
Austropotamobius
recognition features 33-34
species classification 14
pallipes **35, 234**
legal protection 13
status of subspecies 33
torrentium 33
Austropotamobius (Atlantoastacus)
pallipes italicus 34

pallipes pallipes 34
Azeffane 68

Bakhouche 68
Beerkreeft 70, 72
Bernard l'hermite 86
Bernard's hermit crab,
see *Pagurus bernhardus*
Bidenticulate crab 134
Bieberstein's freshwater crab,
see *Potamon ibericum*
Björnkräfta 70, 72
Bjørnkreps 70, 72
Blakrabba 148
Blakrabbe 148
Blaukrabbe 148
Blauwe zwemkrab 148
Blauwpootzweemkrab 142
Blue
crab
restriction of name 148
crawfish 68
-legged swimming crab 142
manna 226
king stone crab,
see *Paralithodes platypus*
swimming crab,
see *Portunus pelagicus*
Boc 172
Bocek 66
Bogavante 56
Boron 152
Box crabs
classification 17
identification 120, 124
Brachynotus sexdentatus
classification 19
immigrant 115
Brachyura
recognition features 23
vernacular name 12
Bristly xanthid crab,
see *Pilumnus hirtellus*
Broad
circular crab,
see *Atelecyclus undecimdentatus*
-clawed porcelain crab,
see *Porcellana platycheles*
Brown king stone crab,
see *Paralithodes brevipes*
Bruja 86
Buey de mar 136
Büyük ayi istakozu 70

Cabra 172
Cabrot 172
Californian crayfish 180
Callianassa subterranea 78
Callianassa tyrrhena **79**
classification 16
recognition features 78
Callinectes sapidus **149, 257**
classification 18
economics 117
in tariffs 177
recognition features 148
Çalpara 146
Cambara 172
Cambarus
affinis 182
clarkii 184
Camber crayfish 182
Cancer bellianus
classification 17
recognition features 136
Cancer pagurus **137, 256**
classification 17
economics 116
legal landed sizes 136
mating 5
recognition features 136
Cangrejo
aterciopelado 156
azul 148
de los mejillones 158
de rio 34
de rio americano 182
de señal 180
dentado 132
moruno 152
vert 150
Cangrexo 150

Cape rock crawfish,
 see *Jasus (Jasus) lalandii*
Caranguejo-mouro 152
Caranguexo 150
Carapace 1
Carcinides maenas 150
Carcinus
 classification 18
 recognition features 150
 aestuari
  economics 118, 150
 maenas **151**
  economics 117
  moulting 5
Caribbean
 spiny crawfish,
  see *Panulirus argus*
 xanthid crab,
  see *Dyspanopeus sayi*
Carraço 140
Casa-alugada 86
Caspian crayfish,
 see *Astacus pachypus*
Cavaco 70
Cavuru 142
Centolla 172
cephalothorax 1
Cervimunida johni **221, 252**
 classification 16
 imported 220
 recognition features 220
Chaceon
 classification 17
 affinis **139**
  compared with *C maritae* 227
  recognition features 138
 maritae **228, 263**
  imported 226
Chancre 150
 de drague 136
Charybdis species
 classification 18
 immigrants 116
Chelipeds 2
Cherax albidus 186
Cherax
 classification 14
 importations 178
 recognition features 186, 188
 albidus
  aquaculture 186
 destructor **187, 240**
  confused taxonomy of 186
 tenuimanus **189, 241**
  recognition features 188
Chesapeake bay swimming crab,
 see *Callinectes sapidus*
Chimbori 226
Chinese
 mitten crab 162
 wolhandkrab 162
Ch'inghsieh 226
Chionoecetes
 classification 19
 importations 118, 226
 in tariffs 177
 recognition features
  bairdi 227
  opilio elongatus 227
  opilio opilio **230, 265**
  tanneri 227
 status of *opilio elongatus* 226
Cicala di mare 72
 a grande 70
Cicaledda 80
Cigala 58
Cigale noire 70
Cigarra 70
 de mar 70
Cingene pavuryasi 150
Circular crabs
 classification 17
 identifications 120, 134
Cirkelronde krab 134
Classification
 changes in 13
 of decapods 13
Clawed lobsters
 fisheries 44-51
 importations 41, 190
 marketing 51-4
 life history 42-3

native species 41
sexing 5
species numbers 21
recognition features 21, 54-5, 190-3
Cleanser crab 142
Clibanarius erythropus **93**
classification 16
recognition features 92
Common
crab 136
hermit crab 86
lobster 56
spider crab 172
Corbola 80
selvatica falsa 78
Corbols salvadega 78
Corystes cassivelaunus **133**
classification 17
recognition features 120, 132
Corwich 172
Crabbe epineux 96
Craba e'mari 172
Crabe
à laine 146
à sardine 140
blue 148
chinoise 162
de roche 96
dormeur 122
enragé 150
fantôme 168
lithodid 96
masqué 132
nageur 144
poilu 156
verruquex 152
vert 150
Cranca 172
Crawfishes
aquaculture 64
classification 15
commercial species 62, 198
fisheries 62, 198
life history 63
native species 62
recognition features 22, 61, 64-5, 199, 202-9
sexing 6
species numbers 22
vernacular names usage 198
Crayfish, *see* freshwater crayfishes
Crevette fouisseuse 80

Demoiselle 146
Demon-faced
crabs
classification 17
recognition features 120
porter crab,
see *Medorippe lanata*
Deniz örümcegi 172
Devil crab 96
Diogenes pugilator **91**
classification 16
recognition features 90
Dohlenkrebs 34
Dromia personata **123**
classification 17
recognition features 122
Dromia vulgaris 122
Dromie 122
Dublin bay prawn 58
Dwarf xanthid crab,
see *Rhithropanopeus harrisii*
Dybvandshummer 58
Dyspanopeus sayi
classification 18
immigrant 115

Ebalia tuberosa **131**
classification 17
recognition features 130
Écrevisse
à patte grêles 38
à pieds blancs 34
de californie 180
rouge de marais 184
Edible
crabs
classification 17

recognition features 120
meat 4
Endophragmal skeleton 4, 5
Erakkorapu 86
Eremitkräfta 86
Eremitkrebs 86
Eremittkreps 86
Eriocheir
classification 19
sinensis **163, 261**
economics 162
importations 188, 226
recognition features 162
status of *japonica* 162
Eriphia spiniforns 152
Eriphia verrucosa **153, 260**
classification 18
economics 118
recognition features 152
Ertekrabbe 158
Erwtenkrabbetje 158
Espanjanhummeri 72
Etrille commune 146
pattes bleues 142
Euastacus armatus
classification 14
size 188
Eucrate crenata
classification
immigrant 116
Eupagurus
bernhardus 86
prideauxi 88
European
clawed lobster,
see *Homarus gammarus*
spiny crawfish,
see *Palinurus elephas*
Evibacus princeps **216**
classification 15
Exoskeleton 3, 4
eyes 1

Face crab 128, 134
False
crabs
recognition features 22
species numbers 22
lobsters
recognition features 22
sexing 5
species numbers 22
Falsa nécora 142
Fan and mitten lobsters
classification 15
importations 215
Fanse krabber 96
Farzit 70
Favollo 152
Fiddler crabs
classification 19
identification 121
Flagella 1
Flathead lobsters
classification 15
Flat lobster 70
Flat-nosed lobster 72
Flodkräfta 34
Fluwele zwemkrab 146
Flying gillie 144
Flying swimming crab,
see *Liocarcinus holsatus*
Formosan clawed lobster,
see *Metanephrops formosanus*
Franquet fuster 128
Freshwater
crabs
classification 19
recognition features 121
crayfishes
classification 13
data base for 179
diseases 31
farming 28-31
introduced species 28
legislation 179
life history 28
native species 27
recognition features 21, 27, 33
species numbers 21
vernacular names 198
Furrowed xanthid crab,

see *Xantho incisus*

Gaddakrabba 96
Galathea glabra 102
Galathea
classification 16
recognition features 101-2, 104
squamifera **103**
strigosa **105**
Galatheidae
vernacular name 12
Galician crayfish 38
Gambero
di fiume 34
di fiume turco 38
Gebochelde spinkrab 174
Gelasimus tangeri 164
Geryon affinis 138
Geryonid crabs
classification 17
recognition features 120-1, 138, 227
Gestekelde sponspootkrabe 168
Gewone
hooiwagenkrab 170
zwemkrab 144
Ghost mud lobster,
see *Callianassa tyrrhena*
Giant mangrove swimming crab 226
Gibb's spider crab,
see *Pisa armata*
Gills 3
Gkagkala 172
Göl istakozu 34
Goneplacid crabs
classification 18
Gonopods 2
Granceola 172
Grancevola 172
Granchio 142
nuotatore 148
Granciporro 136
Grande cigale 70
Granzoporo 152
Great
crab 136
norway crab 96
Green crab 150
Green rock crawfish,
see *Jasus (Sagmariasus) verreauxi*
Grimaldi's stone crab,
see *Neolithodes grimaldii*
Grosse seespinne 172
Grosse cigale 70
Grosser bärenkrebs 70
Grote spinkrab 172
Guernsey crab 136

Hairy
box crab,
see *Homola barbata*
crab 156
-fisted crab 162
Harbour
crab 150
swimming crab,
see *Liocarcinus depurator*
Harig porceleinkrabbetje 110
Havskräft 58
Hãxkrabbe 124
Helmet crab,
see *Corystes cassivaulunus*
Helmkrab 132
Henslow's
schwimmkrabbe 140
swimming crab 140
Heremietkrab 86
Hermit crabs
bionomics 84
classification 16
life history 83
recognition features 22, 83, 85
sexing 5
Heteropanope laevis
classification 18
immigrant 116
Homard 56
Homarus
classification 14
species compared 190-1
americanus **191, 243**
economics 190

recognition features 190
gammarus **57, 191, 242**
largest 56
legal landed sizes 56
recognition features 56, 190
Homola barbata **127**
classification 17
recognition features 126
Hummer 56

Ibacus novemdentatus
figure 216
Import licences 177
Imported species
freshwater crayfishes 178-189
clawed lobsters 190-7
crawfishes 198-214
shovel-nosed lobsters 215-19
squat lobsters 220-1
stone crabs 222-5
true crabs 226-230
Inachus dorsettensis **169**
classification 19
recognition features 168
Intersegmental membranes 2
Isohamahakkirapu 172
Istakos 56
Italian freshwater crab,
see *Potamon fluviatile fluviatile*
Italienischer taschenkrebs 152
Italikos kavouras 148

Jaiba 142
Japanese clawed lobster,
see *Metanephrops japonicus*
Japanese mitten crab,
see *Eriocheir japonica*
Jarradh el bahr 66
Jasus
classification 15
recognition features 210-13
(Jasus) edwardsii **210-11**
(Jasus) lalandii **210-11, 249**
(Jasus) paulensis **210-11**
(Jasus) tristani **210-11**
(Sagmariasus) verreauxi **210-11**
size 198
as tourists trophies 198

Kaiserhummer 58
Karavida 58
tou glykou nerou 34
Kavouras 136, 150
Keisarihummeri 58
Keonid 172
Kerevit 38
Kerry witch 146
Kesis istakozu 88
Khadapi chimbori 226
King crab 172
Kleiner barenkrebs 72
Krabbe 136
Krabbtaska 136
Krank-avel 140
Kreeft 56
Küçük ayi istakozu 72

Lagosta 66
da pedra 72
verde 68
Langoest 66
Langosta
común 66
real 68
Langostim 58
do rio 34
Langostino
amarillo 220
colorado 220
Langouste
rouge 66
royal 68
Langoustine 58
Langspriet 112
Langust 66
Languste 66
Languster 66
Langusti 66
Latreille's hermit crab,
see *Clibanarius erythropus*
Lavagante 56
Leach's squat or plated lobster,

see *Galathea squamifera*
Legal landed sizes
Cancer 136
Homarus 56
lesser spider crab,
see *Maja crispata*
Leucosia signata
classification 17
immigrant 116
Linnaeus's
pea crab,
see *Pinnotheres pisum*
sponge crab,
see *Dromia personata*
Liocarcinus
classification 18
recognition features 142, 144
depurator **143**
holsatus **145**
Liocarcinus puber 146
Lithodes antarctica 222
Lithodes
classification 16
recognition features 96, 222-223
maja **97**
santolla **224, 255**
commerce 222
Lithodidae
vernacular name 12
Little hairy crab 122
Livid swimming crab 144
Long-clawed
crab 132
crayfish,
see *Astacus leptodactylus salinus*
porcelain crab,
see *Pisidia longicornis*
Long-legged spider crab,
see *Macropodia rostrata*
Louisiana swamp crayfish 184
Lyra 72

Macropipus
depurator 142
holsatus 144
Macropode 170
Macropodia rostrata **171**
classification 19
recognition features 170
Madeiran crab,
see *Cancer bellianus*
Magnosa 70
Maja
classification 19
crispata
vernacular names 172
squinado **173, 264**
confusion with stone crabs 95
economics 117, 172
recognition features 172
Mamun 80
Mandibles 1
Mangrove crab 226
Marbled rock crab,
see *Pachygrapsus marmoratus*
Marine crayfishes
vernacular name 198
Marron,
see *Cherax tenuimanus*
Marsh crayfish 38
Mary crab 142
Masineta 142
Masked crab 132
Maskeringskrabba 172
Mating 5
Maulwurfkrebs 78, 80
Mavi Yengeç 148
Maxillae 1
Maxillipeds 2
Mediterranean
mud lobster,
see *Upogebia pusilla*
rock crab,
see *Brachynotus sexdentatus*
slipper lobster,
see *Scyllarides latus*
shore crab,
see *Carcinus aestuari*
Medorippe lanata **129**
classification 17
recognition features 128
Melbourne Yabby,

see *Cherax destructor*
Metanephrops **195**
classification 14
cf norway lobster 192
fisheries 194
species compared 196-7
andamanicus **193,** 196, **245**
armatus, 196
formosanus 196
japonicus 196
mozambicus 197
sagamiensis 197
thomsoni 197
velutinus 197
Mitten crab,
see *Eriocheir sinensis*
Montagu's crab 154
Moreton bay flathead lobster,
see *Thenus orientalis*
Moroccan fiddler crab,
see *Uca tangeri*
Morphology *x*
Mosseelkrabbetje 158
Moulting 5
Mouthparts 1
Mud
lobsters
bionomics 76
classification 16
life history 75
recognition features 22, 76
sexing 6
species numbers 22
tunnels of **77**
shrimps 78, 80
spiny crawfish,
see *Panulirus polyphagus*
swimming crab,
see *Scylla serrata*
Munida bamffica 106
Munida rugosa **107**
classification 16
recognition features 106
Murray cray,
see *Euastacus armatus*
Muscles 4
Myra fugax
classification 17
immigrant 116

Narrow-mouth crabs
classification 17
Navalheira 146
-azul 148
Nécora 146
Necora puber **147, 257**
classification 18
economics 117
recognition features 146
Nefrops 58
Neolithodes grimaldii
classification 16
recognition features 95
Nephropoidea
vernacular name 12
Nephrops norvegicus **59, 193, 244**
classification 15
legal landed sizes 58
recognition features 58, 192
Nephropsis atlantica
classification 15
recognition features 55
Nili chimbori 226
Nipper crab 140
Noble crayfish,
see *Astacus astacus astacus*
Nomenclature 11
Nona kankra 226
Noordsee
-einsiedlerkrebs 86
-taschenkrebs 136
Noordzeekrab 136
Noorse kreeft 58
North atlantic shore crab,
see *Carcinus maenas*
Northern stone crab,
see *Lithodes maia*
Norway
clawed lobster,
see *Nephrops norvegicus*
crab 96
Nurrabeil yabby,

see *Cherax albidus*
Nut crabs
classification 17
recognition features 120

Old man's face crab 132
oosetae 3
Orconectes limosus 182
Orconectes (Faxonius)
limosus **183, 238**
classification 14
commercial value 178
recognition features 182
Ornate spiny lobster,
see *Panulirus ornatus*
Ox crab 136

Pachygrapsus marmoratus **161**
classification 19
recognition features 160
Pacifastacus leniusculus 180
Pacifastacus (Pacifastacus)
leniusculus leniusculus **181, 237**
classification 14
farming 29
importations 178
recognition features 237
Pacific snow spider crab,
see Chionoecetes opilio
elongatus
Paguridae
vernacular name 12
Paguro bernado 88
Pagurus
classification 16
recognition features 86, 88
bernhardus **87**
prideaux **89**
Pagurus solitarius 88
Painted spiny crawfish,
see *Panulirus versicolor*
Palinuridae
vernacular names 12, 198
genera compared 202
Palinurus
classification 15
recognition features 66, 202
elephas **67, 200-1, 246**
mauritanicus **67, 247**
Panulirus
classification 15
fisheries 198
in tariffs 177
marketing 199
recognition features 68, 203, 208-9
vernacular names 198
argus **204, 206**
cygnus **204, 206**
homarus **205, 207**
ornatus **204, 206**
polyphagus **205, 207**
regius **69, 248**
versicolor **205, 207**
Parachaeraps bicarinatus 186
Paralithodes
classification 16
recognition features
brevipes 222-3
camtschaticus 222, **225, 255**
in tariffs 117
fisheries 222
platypus 223
Paralomis
classification 16
recognition features 95
bouvieri 95
spectabilis 95
Paramola cuvieri **125**
classification 17
recognition features 124
Paromole 123
Parribacus antarcticus **216**
classification 15
commerce 215
Patagonian royal stone crab,
see *Lithodes santolla*
Pateixo 140
Patelo 140
Paura 152
Pavurya 152
Pea crabs
classification 18

recognition features 121
Pennant's
ebalia 130
nut crab,
see *Ebalia tuberosa*
spider crab 168
Petite cigale 72
Petrokavouras 152
Pilado 140
Pilumnoides inglei
classification 18
introduced 115
Pilumnopeus vauquelini
classification 18
immigrant 116
Pilumnus hirtellus **157**
classification 18
recognition features 156
Pinnotheré 158
Pinnotheres
subspecies status 158
pisum **159**
classification 18
recognition features 158
Pisa armata **175**
classification 19
recognition features 174
Pisidia
classification 17
recognition features
bluteli 109
longicornis 112, **113**
longimana 109
Plated lobster 102, 106
Pleopods 2, 3
Pleuroncodes
classification 17
recognition features 220
monodon **221, 254**
imported 220
planipes status of 220
Plomb 172
Poinclos 136
Polybius henslowii **141**
classification 18
recognition features 140
Pond crayfish 38
Porcelaenskrabbe 110
Porcelain crabs
classification 17
bionomics 109
life history 108
recognition features 99, 109
Porceleinkrabbetje 112
Porcellana longicornis 112
Porcellana platycheles **111**
classification 17
recognition features 110
Porcellanidae
vernacular names 12
Porselenskrabber 110
Porter crab 128
Portunus
depurator 142
holsatus 144
puber 146
Portunus pelagicus **229, 258**
classification 18
immigrant 116
importations 118, 226
recognition features 227
Potamobius
fluviatilis 36
leniusculus 180
Potamon
classification 19
recognition features 166
fluviatile fluviatile **167**
ibericum **167**
potamios **167**
Prickly crab 96
Prideaux's hermit crab 88,
see *Pagurus prideaux*
Procambarus clarkii 184
Procambarus (Scapulicambarus)
clarkii **185, 239**
classification 14
damage caused by 178
recognition features 184
Puerulus sewelli **200-1, 214, 252**
classification 15
recognition features 202

Pygmy locust lobster,
see *Scyllarus pygmaeus*

Radjungan 226
Rantataskurapu 150
Rauhe schwimmkrabbe 142
Red
-banded clawed lobster,
see *Metanephrops thomsoni*
crab 220
crayfish 184
geryonid crab,
see *Chaceon affinis*
lobster 220
rock crayfish,
see *Jasus (Sagmariasus) edwardsii*
slipper lobster,
see *Scyllarides herklotsii*
squat lobster,
see *Pleuroncodes monodon*
swamp crayfish,
see *Procambarus (Scapulicambarus) clarkii*
Regulations
food labelling 13
Rhithropanopeus harrisii
classification 18
immigrant 118
Risso's box crab,
see *Paromola cuvieri*
Rivierkreeft 34
Rock
crab(s) 160
classification 19
recognition features 121
crawfish
name restricted to *Jasus* 198
marketing 199
recognition features 210-13
lobster 66
vernacular name usage 198
Rostrum 1
Roux's hermit crab,
see *Diogenes pugilator*
Royal spanish lobster,
see *Arctides regalis*
Royal spiny
crawfish,
see *Panulirus regius*
lobster 68
Rugose squat lobster,
see *Munida rugosa*
Ruig krabbetje 156
Running crabs
classification 19
recognition features 121
Ruwe kiezelkrab 130

St Paul Rock lobster,
see *Jasus (Jasus) paulensis*
Samoan crab 226
Samtkrabbe 146
Sand
crab 226
ghost lobster 72, 78
ghost shrimp 72
Sandkrebs 78
Santiaguito 72
Santola 172
Sapateira 136
Sardine swimming crab,
see *Polybius henslowii*
Sastre negro 102
Scalloped spiny crawfish,
see *Panulirus homarus*
Scampi 13, 58
Scampo 58
Scaphognathite 1
Scarlet clawed lobster,
see *Nephropsis atlantica*
Scorpion spider crab,
see *Inachus dorsettensis*
Scylla serrata **229, 259**
classification 18
importations 118, 226
recognition features 227
Sculptured
clawed lobster,
see *Metanephrops sagamiensis*
mitten lobster,
see *Parribacus antarcticus*

Scyllaridae
vernacular name 12
Scyllarides
classification 15
recognition features 70, 219
herklotsii **217**
latus **71, 250**
cf imported species 215
Scyllarus
cf imported species 215
recognition features 72, 219
arctus **73, 251**
batei **217**
pygmaeus 72
Sexing decapods 2, 3, 5
Shield 3
Shield fan lobster,
see *Evibacus princeps*
Shore crabs
classification 18
recognition features 121
Shovel-nosed lobsters
classification 15
imported species 215
life history 63
species numbers 22
recognition features 22, 61, 64, 215, 218-19
Siderokavouras 146
Signal crayfish,
see *Pacifastacus (Pacifastacus) leniusculus leniusculus*
Signalkrebse 180
Simkrabba 142, 146
Simontorps crayfish 180
Sinitaskurapu 148
Sjøkreps 58
Skaltsini 88
Sleepy crab 122
Slender-legged spider crab 170
Small european locust lobster,
see *Scyllarus arctus*
Smooth fan lobster,
see *Ibacus novemdentatus*
Snow
queen spider crab,
see *Chionoecetes bairdi*
spider crabs
recognition features 227
Soft locust lobster,
see *Scyllarus batei*
Soldier crab 86
Soum 68
Spanish and slipper lobsters
classification 15
imported species 215
recognition features 72, 219
Sperm recaptacle 2
Spider crabs
classification 19
recognition features 121
Spindelkrabba 172
Spiny
crawfishes and lobsters,
restriction of name 198
lobster 66
spider crab 172
squat lobster 104
Sponge crabs
classification 17
recognition features 120
Squat lobsters
classification 16
bionomics 101
commerce 100
imported species 220
life history 100
recognition features 23, 99, 101
Stankelbenskrabbe 168, 170
Steinkrabbe 96
Stenorhynchus rostratus 170
Sternum 3
Stone
crab(s) 226
classification 16
commerce 95
imported species 222
life history 94
recognition features 95, 222
crayfish,
see *Austropotamobius*

torrentium
Strandkrab 150
Strandkrabbe 150
Strigose squat lobster,
see *Galathea strigosa*
Striped crayfish,
see *Orconectes (Faxonius) limosus*
Sumpfkrebs 38
Suømmekrabber 144
Svømmekrabbe 142, 146
Swamp
crab 226
crayfish 38
Swimming crabs
classification 18
recognition features 121

Tanner spider crab,
see *Chionoecetes tanneri*
Taskekrabbe 136
Tasmanian crayfish,
see *Astacopsis gouldi*
Taxonomy 11
Telson 1, 3
Thalamita poissonii
classification 18
immigrant 116
Thalassina littoralis 80
Thalassinidea
vernacular name 12
Thelxiope barbata 126
Thenus orientalis **217**
classification 15
commerce 215
Thoracic limbs 2
Thorn-back spider crab,
see *Maja squinado*
Topo de mar 78
Tourteau poupart 136
Tristan rock crawfish,
see *Jasus (Jasus) tristani*
Troldhummer 102
Troldkrabbe 172
Trolkrabber 96
Trollhummer 106
True crabs
classification 17
imported species 226
introduced species 115
life history 118
life styles 119
native species fished 116
recognition features 23, 120
sexing 6
species numbers 23
Turkish crayfish 38

Uca tangeri **164, 262**
classification 19
economics 164
recognition features 164
Uneven crab 130
UK custom and excise tariff 177
Upogebia littoralis 80
Upogebia pusilla **81**
classification 16
recognition features 80
Uropods 3

Varunid crabs
classification 19
Velvet
clawed lobster,
see *Metanephrops velutinus*
fiddler crab 146
swimming crab,
see *Necora puber*
Vernacular names *x*, 12, 198

Warty xanthid crab,
see *Eriphia verrucosa*
West african
fiddler crab 164
geryonid crab,
see *Chaceon maritae*
White
-clawed crayfish,
see *Austropotamobius pallipes*
-footed crayfish 34
Whip crawfishes
restriction of name 198

importations 199
Wolhandkrabben 162
Wolkrab 122

Xanthid crabs
classification 18
recognition features 121
Xantho floridus 154
Xantho
classification 18
recognition features 154
incisus **155**
pilipes 154

Yellow squat lobster,
see *Cervmunida johni*
Yengeç 140, 142

Zespoot 112
Ziz el bahr 72
Zwemkrab 142, 144

MIX
Papier aus verantwortungsvollen Quellen
Paper from responsible sources
FSC® C105338

If you have any concerns about our products,
you can contact us on
ProductSafety@springernature.com

In case Publisher is established outside the EU,
the EU authorized representative is:
Springer Nature Customer Service Center GmbH
Europaplatz 3, 69115 Heidelberg, Germany

Printed by Libri Plureos GmbH
in Hamburg, Germany